Anisotropy Effects in Superconductors

Anisotropy Effects in Superconductors

Edited by
Harald W. Weber
Atomic Institute of the Austrian Universities
Vienna, Austria

Plenum Press · New York and London

Library of Congress Cataloging in Publication Data

Main entry under title:

Anisotropy effects in superconductors.

"Proceedings of an international discussion meeting held at Atominstitut der Österreichischen Universitäten, Vienna, Austria, April 21-23, 1976."
Includes indexes.
1. Superconductors—Congresses. 2. Anisotropy—Congresses. I. Weber, Harald W.
QC612.S8A54   537.6'23   76-56737
ISBN 0-306-31006-6

Proceedings of an International Discussion Meeting held at
Atominstitut der Österreichischen Universitäten, Vienna, Austria, April 21-23, 1976

© 1977 Plenum Press, New York
A Division of Plenum Publishing Corporation
227 West 17th Street, New York, N.Y. 10011

All rights reserved

No part of this book may be reproduced, stored in a retrieval system, or transmitted, in any form or by any means, electronic, mechanical, photocopying, microfilming, recording, or otherwise, without written permission from the Publisher

Printed in the United States of America

PREFACE

Considering the enormous progress achieved in some special areas of superconductivity during the last few years, it seemed worthwhile to discuss thoroughly a subject, which has encountered severe problems on the theoretical and the experimental side, namely the effects of the anisotropic electron and phonon properties of (single crystalline) materials on the characteristic features of the superconducting state.

The fact that the majority of scientists actively engaged in this research field at present were brought together at the meeting, has led to an almost complete coverage of the pertinent topics in this volume. Six review papers discuss the development and the present state of theory and experiment concerning the anisotropies of the upper critical field and the magnetization as well as the flux line lattice and the superconducting energy gap. Furthermore, 18 papers present the most recent research on these properties and additional anisotropy effects associated with the intermediate state patterns or the specific heat. Concerning the magnetic properties, remarkable agreement between theory and experiment has been achieved; on the other hand, controversial views regarding the existence of anisotropy in the superconducting energy gap are presented. In both cases, however, theorists and experimentalists are still confronted with a number of open questions. It is the purpose of this book to draw the attention of scientists to these fascinating problems.

Vienna, July 1976   Harald W. Weber

ACKNOWLEDGEMENTS

I wish to express my appreciation to the Directors of the Institute, Prof.Dr.G.Eder and Prof.Dr.H.Rauch, for making the facilities of the Institute available for the meeting. Furthermore, the help of graduate students and staff members is acknowledged.

I am particularly grateful for the financial support received from Bundesministerium für Wissenschaft und Forschung, Stadt Wien, Technische Universität Wien and Atominstitut der Österreichischen Universitäten.

Furthermore, I wish to thank all participants of the meeting and in particular all authors for the prompt submission of their contributions.

Finally, I wish to thank Miss Eva Haberl for her help with the preparation of the meeting and the excellent typing of this book.

H.W.W.

CONTENTS

1. THE UPPER CRITICAL FIELD

R-1 H.TEICHLER (invited):
     On the Theory of Macroscopic Anisotropy Phenomena
     in Type-II Superconductors 7

R-2 T.OHTSUKA (invited):
     Experimental Investigations on the Anisotropy of the
     Upper Critical Field in Type-II Superconductors 27

2. H_{c2} AND RELATED PROPERTIES

C-1 P.ENTEL and M.PETER:
     The Influence of Fermi Surface Anisotropy on $H_{c2}(T)$ ... 47

C-2 E.SEIDL and H.W.WEBER:
     Impurity Dependence of H_{c2} Anisotropy in Niobium 57

C-3 H.TEICHLER:
     Microscopic Anisotropy Parameters of Niobium 65

C-4 H.KIESSIG, U.ESSMANN, H.TEICHLER and W.WIETHAUP:
     Anisotropy of H_{c2} in PbTl-Alloys 69

C-5 K.TAKANAKA:
     Comments on H_{c2} at Low Temperatures 75

C-6 C.E.GOUGH:
     The Influence of Magnetic Anisotropy on the Properties
     of Niobium in the Mixed State 79

C-7 H.R.OTT:
     Anisotropy of the Stress Dependence of Critical Parameters in Uniaxial Superconductors 87

3. FLUX LINES

R-3 K.TAKANAKA (invited):
Magnetization and Flux Line Lattice in Anisotropic
Superconductors 93

R-4 J.SCHELTEN (invited):
Morphology of Flux Line Lattices in Single-Crystalline
Type-II Superconductors 113

4. FLUX LINES, DOMAINS AND MAGNETIZATION

C-8 B.OBST:
Correlations between Flux Line Lattice and Crystal
Lattice 139

C-9 W.RODEWALD:
Experiments on the Correlation between the Flux Line
Lattice and the Crystal Lattice in Superconducting
Lead Films 159

C-10 R.P.HUEBENER, R.T.KAMPWIRTH and D.E.FARRELL:
Anisotropy in the Intermediate State of Superconducting
Mercury 165

C-11 F.K.MULLEN, R.J.HEMBACH and R.W.GENBERG:
Mixed State Anisotropy of Superconducting Vanadium 171

C-12 R.SCHNEIDER, J.SCHELTEN and C.HEIDEN:
Measurement of Torque due to Anisotropy of the Magnetization Vector in Superconducting Niobium 177

5. ENERGY GAP

R-5 J.P.CARBOTTE (invited):
Microscopic Calculations of Energy Gap Anisotropy 183

Contents

R-6   J.L.BOSTOCK and M.L.A. MAC VICAR (invited):
An Evaluation of the Validity of Superconducting
Evidence for Anisotropy and Multiple Energy Gaps 213

6. ENERGY GAP AND RELATED PROPERTIES

C-13   M.L.A. MAC VICAR, J.L.BOSTOCK and K.R.MILKOVE:
Tunneling Junction Phenomena: An Answer to Unanswered
Questions 257

C-14   W.D.GREGORY, A.J.GREKAS, S.HORN and L.MORELLI:
An Analysis of Evidence for Superconducting Energy Gap
and Pairing Interaction Anisotropy for Two Types of
Experiments 265

C-15   N.C.CIRILLO, Jr. and W.L.CLINTON:
A Nearly Free Electron Model of the BCS Gap Equation:
Energy Gap Anisotropy in Gallium 283

C-16   R.W.STARK and S.AULUCK:
Specific Heat of Superconducting Zinc 293

7. OTHER TOPICS

C-17   J.M.DUPART, J.ROSENBLATT and J.BAIXERAS:
Unusual Resistance Effect Shown in a Periodic S-N-S
System (Pb-Sn Lamellar Eutectic) 299

C-18   L.J.AZEVEDO, W.G.CLARK, G.DEUTSCHER, R.L.GREENE,
G.B.STREET and L.J.SUTER:
The Upper Critical Field of Superconducting Polysulfur
Nitride, $(SN)_x$ - (Abstract only) 307

List of Participants 309
Author Index 313
Subject Index 315

R-1   ON THE THEORY OF MACROSCOPIC ANISOTROPY PHENOMENA IN TYPE-II SUPERCONDUCTORS

H. Teichler

Institut für Physik am Max-Planck-Institut für Metallforschung and Institut für Theoretische und Angewandte Physik der Universität, D-7 Stuttgart, W-Germany

Abstract

The interrelationships between macroscopic anisotropy phenomena in the mixed state of monocrystalline type-II superconductors and the microscopic anisotropies of the superconducting electron system are discussed. In particular a recent theory of H_{c2} anisotropy in metals with low or moderate impurity content at arbitrary temperatures below T_c is considered. According to this theory the anisotropy of the energy gap of the pure material and details of the electron band structure anisotropy may be deduced from investigations of the temperature and impurity dependence of H_{c2} anisotropy. In addition, a nonlocal theory of preferential orientations of the flux line lattice relative to the crystal lattice is presented.

1. Introduction

The general topic to be considered in this review is the inter-

relationship between macroscopic anisotropy phenomena in the mixed state of monocrystalline type-II superconductors and the microscopic anisotropies of the superconducting electron system. Here "macroscopic anisotropy phenomena" (MAP's) comprise all those anisotropies of superconductors that are due to spatial variations in the system on a scale large compared to the crystal lattice constant. At present, in particular MAP's in the magnetization curve and in the flux line arrangements are studied. MAP's observed in the magnetization curves are, e.g.,
- the anisotropy of the upper critical field H_{c2} (e.g. /1-3/),
- the anisotropy of κ_2 /4-6/,
- the anisotropy of H_{c1} /7,8/ and of the spontaneous induction B_o at H_{c1} /7/.

(There may even exist an anisotropy of H_{c3}.) Furthermore, by direct observation of the structure of the flux line lattice using the decoration technique of Träuble and Essmann /9/ or by neutron diffraction experiments /10/ one finds
- correlations between the symmetry of the crystal lattice and the structure of the FLL. E.g. cubic Nb samples with the magnetic field direction parallel to the fourfold <100> crystallographic direction have a square FLL /11/ (as shown schematically in Fig.1), whereas the FLL is triangular for magnetic fields parallel to the <111> axis;
- preferential orientations of the basis vectors of the FLL relative to the crystal axes /10-12/ (also shown schematically in Fig.1).

The MAP's reflect the microscopic anisotropies of the superconducting electron system, i.e., the anisotropy of the electron band structure of the normal state (anisotropy of the shape of the Fermi surface, of the Fermi velocity and of the density of states at the Fermi surface) and the anisotropy of the attractive interaction (including the electron-phonon interaction, phonon dispersion curves and Coulomb pseudopotential). Effects of these microscopic anisotropies on the magnetization curve or the flux line arrangement may be

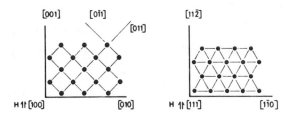

Fig.1: Schematic plot of flux line arrangements in monocrystalline Nb samples with different crystallographic orientations

investigated on various levels of theory, ranging from anisotropic local Ginzburg-Landau (GL) equations /13,14/ to anisotropic generalizations /15/ of the highly nonlocal Gor'kov theory /16/. In order to present the different stages of theoretical treatment of the MAP's in a clear and concise manner we mainly consider in the following the anisotropy of H_{c2}. In addition we discuss briefly a nonlocal theory dealing with the problem of preferential orientations of the FLL relative to the CL. The theoretical treatment of this phenomena and of further MAP's in the framework of anisotropic GL equations /13,14/ and generalizations /6,17/ of these equations including some nonlocal corrections of Gor'kov's theory will be reviewed in another paper /18/.

2. Theory of H_{c2} Anisotropy in Uniaxial Superconductors

The basic features of H_{c2} anisotropy in uniaxial systems like tetragonal or hexagonal metals or layered compounds may be evaluated within the framework of the GL theory, whereas this theory completely fails to describe the H_{c2} anisotropy in cubic systems /19/. Due to this fact we shall discuss these two classes of systems separately.

The GL theory originally has been formulated for isotropic model systems, but anisotropic generalizations of this theory accounting for the existence of MAP's have been already derived in 1963 by Gor'kov and Melik-Barkhudarov /13/ and by Caroli et al. /14/. Within these anisotropic theories the GL free energy functional is given by

$$\Delta\Omega_{SN} = \int d^3\underline{r}\, \psi^*(\underline{r}) \{ \frac{1}{4} \sum_{i,j} (m^{-1})_{ij} \hat{p}_i \hat{p}_j - |\alpha| + \frac{1}{2}\beta|\psi(\underline{r})|^2 \} \psi(\underline{r})$$
$$+ \frac{1}{4\pi} \int d^3\underline{r}(\underline{H} - \underline{H}_{ext})^2, \qquad (1)$$

$$\hat{p} = \frac{\hbar}{i}\nabla - \frac{2e}{c}\underline{A}(\underline{r}).$$

Here the effects of microscopic anisotropies are expressed by the anisotropy of the effective mass tensor $(m^{-1})_{ij}$ (which reduces to an isotropic tensor in the case of cubic systems /19/). This effective mass tensor should be distinguished clearly from the band structure effective mass: the quantities $(m^{-1})_{ij}$ are given by

$$(m^{-1})_{ij} = \frac{1}{2\varepsilon_F} < \varphi^2(\underline{k}_F) v_i(\underline{k}_F) v_j(\underline{k}_F) >, \qquad (2)$$

where the brackets denote averages over the Fermi surface

$$<\ldots> = \int_{F.S.} d^2\underline{k}_F\, v^{-1}(\underline{k}_F) \ldots / \int_{F.S.} d^2\underline{k}_F\, v^{-1}(\underline{k}_F), \qquad (3)$$

and

$$\varphi(\underline{k}_F) = \phi(\underline{k}_F)/\sqrt{<\phi^2>} \qquad (4)$$

describes the anisotropy of the Meissner state gap parameter ϕ /13/. Thus $(m^{-1})_{ij}$ includes band structure anisotropies as well as the anisotropies of the attractive interaction, which are incorporated in the anisotropy of the energy gap. Even in the case of layered compounds, where the electron states in adjacent layers are merely coupled by Josephson tunneling, the different types of electron propagation parallel and perpendicular to the layers may be described by an anisotropic effective mass tensor /20-22/.

For uniaxial superconductors at temperatures near T_c the anisotropic GL theory predicts that H_{c2} depends on the orientation of the magnetic field relative to the crystallographic c-axis /23/

$$\frac{e}{c} H_{c2} \equiv \frac{e}{c} H_{c2}(\theta) = \frac{3(\pi k_B T_c)^2}{\langle v_F^2 \rangle \lambda(3)} \cdot \frac{(1 - T/T_c)}{\sqrt{\sin^2\theta + (m_\perp/m_\parallel)\cos^2\theta}} \quad (5)$$

$$\lambda(n) = \sum_{k=0}^{\infty} (2k+1)^{-n}.$$

In (5) the quantities $m_\parallel$, $m_\perp$ denote the effective mass components parallel and perpendicular to the c-axis, and θ denotes the angle between the field direction and the c-axis. In the basal plane no angular dependence of H_{c2} should occur. Moreover, according to (5), the ratio of the upper critical field parallel, $H_{c2\parallel}$, and perpendicular to the c-axis, $H_{c2\perp}$, should be temperature independent

$$H_{c2\parallel}/H_{c2\perp} = \sqrt{m_\perp/m_\parallel} \quad (6)$$

The anisotropy of H_{c2} at temperatures near T_c has been observed on hexagonal Tc by Kostorz et al. /3/. However, in contrast to the predictions of the GL theory Kostorz et al. /3/ found a small increase of $H_{c2\parallel}/H_{c2\perp}$ with decreasing temperatures. Such a temperature dependence violates (6) and is due to nonlocal corrections of Gor'kov's theory to the GL equations. Theories of H_{c2} anisotropy in uniaxial superconductors being capable to describe this temperature dependence have been derived by Takanaka /24/ for temperatures near the transition temperature and by Teichler /25/ for all temperatures below T_c. These theories show, in particular, that the temperature dependence of $H_{c2\parallel}/H_{c2\perp}$ cannot be attributed to an anisotropic electron mass tensor: the T-dependence has to be ascribed to an anisotropic energy gap /24,25/ or, likewise, to a wave vector dependent effective mass of the electrons /25/.

In addition, theory predicts /25/ that H_{c2} shows an angular dependence even in the basal plane for temperatures below T_c. This anisotropy has not yet been detected on Tc. Recently, however, Skokan et al. /26/ observed such an anisotropy of H_{c2} on hexagonal $Cs_{0.1}WO_{2.9}F_{0.1}$, where the magnitude of the relative H_{c2} anisotropy in the basal plane ranged from 0.113 to 0.313 for various samples at a reduced temperature $T/T_c \approx 0.87$.

The theories of H_{c2} anisotropy for uniaxial superconductors in the nonlocal region markedly below T_c are in their basic features similar to the corresponding theories for cubic superconductors. With regard to the fact that nonlocal effects are more important in cubic systems (in so far as they are the only source of MAP's in these systems) we shall close the discussion of uniaxial superconductors at this stage and consider details of the nonlocal theories for the case of cubic materials.

3. Theory of H_{c2} Anisotropy in Cubic Metals

In cubic metals the dependence of H_{c2} on the orientation of the field direction $\underline{e}_H$ relative to the crystal axes obviously has to show cubic symmetry. Hence the angular dependence of $H_{c2}(\underline{e}_H)$ (or the relative anisotropy $\Delta H_{c2}(\underline{e}_H)/\overline{H}_{c2}$) may be expressed in terms of orthonormalized cubic harmonics /27/,

$$\frac{\Delta H_{c2}(\underline{e}_H)}{\overline{H}_{c2}} \equiv \frac{H_{c2}(\underline{e}_H) - \overline{H}_{c2}}{\overline{H}_{c2}} = \sum_{l=4,6,\ldots} H_l(\underline{e}_H) \cdot a_l(T,\tau) \qquad (7)$$

Here $\overline{H}_{c2}$ denotes the average of H_{c2} over all space directions. The cubic harmonics $H_l(\underline{e}_H)$ are those linear combinations of spherical harmonics with angular momentum l that have full cubic symmetry*. The lowest orthonormalized cubic harmonics belong to $l = 0, 4, 6, \ldots$

$$H_0(\underline{e}_H) = 1,$$

$$H_4(\underline{e}_H) = \sqrt{\frac{525}{16}} \, (\beta_1^4 + \beta_2^4 + \beta_3^4 - \frac{3}{5}),$$

$$H_6(\underline{e}_H) = \frac{11 \times 21}{8} \sqrt{26} \, (\beta_1^2 \beta_2^2 \beta_3^2 + \frac{1}{22}(\beta_1^4 + \beta_2^4 + \beta_3^4 - \frac{3}{5}) - \frac{1}{105}),$$

where the β_i's denote the direction cosines of $\underline{e}_H$ with respect to the cubic crystal axes.

The expansion coefficients a_l depend on the microscopic anisotropies of the superconducting electron system and, in addition, on temperature T and impurity content in the sample which may be characterized by the s-wave scattering time τ. The detailed form of these coefficients has to be derived from microscopic theories. For temperatures near T_c and for samples with a sufficiently high impurity content, Hohenberg and Werthamer /19/ proved in 1967 that a nonvanishing coefficient $a_{l=4}$ may result from GL-like equations including the leading nonlocal corrections of Gor'kov's theory (i.e. from an anisotropic Neumann-Tewordt formalism, cf., e.g. /17/). In particular, attributing the H_{c2} anisotropy entirely to electron band structure anisotropy they showed, that $a_{l=4}$ could be factorized into a universal weight function $G(T/T_c;\tau)$ depending on temperature and impurity content and a material parameter describing the anisotropy of Fermi surface and Fermi velocity. Thus the angular dependence of H_{c2} should be of the form

$$\frac{\Delta H_{c2}(\underline{e}_H)}{\overline{H}_{c2}} \sim G(T/T_c;\tau) \, H_4(\underline{e}_H). \tag{8}$$

---

*An additional subscript j should be introduced in the notation of cubic harmonics in order to label the different irreducible representations for one particular l. But since for $l \leq 10$ only one irreducible representation exists at most for each l and since higher angular momentum contributions are far beyond the resolution of present H_{c2} anisotropy measurements, we shall suppress this subscript for convenience.

The experiments of Reed et al. /28/ on Nb showed a more complicated angular variation of H_{c2}. To account for this observation various attempts /6,29-32/ have been made to include higher order nonlocal corrections and to evaluate the higher angular momentum components of H_{c2}. The ranges of validity of these theories depend on the reduced temperature

$$t = T/T_c$$

and on the impurity parameter

$$\alpha = \frac{\hbar}{\tau \, 2\pi \, k_B T_c} \, .$$

The earlier theories /6,29,30/ ascribe the H_{c2} anisotropy entirely to Fermi surface anisotropies. Moreover, they take into account only first and second order nonlocal corrections of Gor'kov's theory to the GL equations. (Because of this fact they will be called "quasi-local" theories in the following.) As indicated in Fig.2 these theories are valid only in a narrow temperature range below T_c. In the (t,α)-plane the boundary of this so-called region of small nonlocality may be described by the implicit relationship /19/

$$\frac{3}{8} \frac{\overline{H_{c2}}(t,\alpha)}{-\frac{\partial}{\partial t} H_{c2}(t,\alpha=0)\Big|_{t=1}} \cdot \frac{1}{(1+\alpha/t)^2} = \text{const.} \ll 1$$

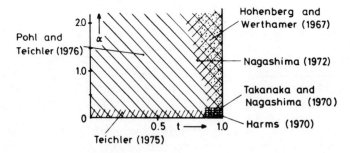

Fig.2: Ranges of validity of various theories of H_{c2} anisotropy in cubic superconductors

which indicates, that the width of this region enlargens with growing α. Since the magnitude of relative H_{c2} anisotropy decreases with increasing temperature and impurity content, the experimental investigations of H_{c2} anisotropy are usually carried out in the region of strong nonlocality, i.e. in metals with low or moderate impurity content and at temperatures clearly below T_c. Because of this situation we shall discuss in the following a recent "nonlocal" theory /31,32/ of H_{c2} anisotropy in more detail, which takes into account all orders of nonlocal corrections and is, therefore, valid even in this region of strong nonlocality.

4. Nonlocal Theory of H_{c2} Anisotropy

The nonlocal theory /31,32/ of H_{c2} anisotropy is based on a description of the superconducting electron-phonon system by means of anisotropic transport-like equations /17/, which represent generalizations of the transport-like equations of, e.g., Eilenberger /33/ or Betbeder-Matibet and Nozieres /34/. Within this theory the anisotropies of the attractive electron-electron interaction as well as those of the electron band structure are considered as sources of H_{c2} anisotropy. In order to derive explicit results on the temperature and impurity dependence of $\Delta H_{c2}/\overline{H}_{c2}$ some simplifications have been introduced in the theory. In particular, the anisotropic attractive electron-electron interaction, which depends in general on the wave vectors of the interacting electrons and on their energies, is replaced by an energy independent and separable model interaction as introduced by Markowitz and Kadanoff /35/

$$V(\underline{k}_F,\underline{k}'_F;\omega,\omega') \rightarrow \varphi(\underline{k}_F) V(0) \varphi(\underline{k}'_F) . \tag{9}$$

Application of this model interaction to the homogeneous Meissner state yields, that $V(0)$ denotes the effective coupling strength which determines T_c, and that $\varphi(\underline{k}_F)$, in accordance with (4), describes the reduced anisotropy of the energy gap of the pure metal.

For impure metals the anisotropy incorporated in $\varphi(\underline{k}_F)$ must be distinguished from the anisotropy of $\omega_o(\underline{k}_F)$, the energy gap in the renormalized excitation spectrum of the superconductor as measured by tunneling experiments /36/. Whereas impurities strongly reduce the anisotropy of $\omega_o(\underline{k}_F)$ /36,37/, $\varphi(\underline{k}_F)$ retains its clean metal structure. (Nevertheless the effect of $\varphi(\underline{k}_F)$ on the anisotropy of H_{c2} is also reduced by impurities.) As an additional approximation only linear effects of the microscopic anisotropies are taken into account, i.e. only linear terms in

$$\delta\varphi^2 = \varphi^2(\underline{k}_F) - 1 \equiv \left.\frac{\varphi^2(\underline{k}_F) - <\varphi^2>}{<\varphi^2>}\right|_{\text{clean limit}}, \qquad (10)$$

$$\frac{\delta v^2}{<v_F^2>} = \frac{v^2(\underline{k}_F) - <v_F^2>}{<v_F^2>}, \qquad (11)$$

are considered. Using these approximations the coefficient $a_l(T,\tau)$ for each angular momentum value l may be represented by /32/

$$a_l(T,\tau) = c_l\{h_l^{(1F)}(t,\alpha)\gamma_l^{(1F)} + h_l^{(1\varphi)}(t,\alpha)\gamma_l^{(1\varphi)} + h_l^{(2)}(t,\alpha)\gamma_l^{(2)}\} \qquad (12)$$

The quantities

$$\gamma_l^{(1F)} = <H_l(\underline{v}_F)>, \qquad (13)$$

$$\gamma_l^{(1\varphi)} = <\delta\varphi^2 H_l(\underline{v}_F)>, \qquad (14)$$

$$\gamma_l^{(2)} = <\frac{\delta v^2}{<v_F^2>} H_l(\underline{v}_F)>, \qquad (15)$$

are material parameters of the host metal: The parameters $\gamma_l^{(1\varphi)}$ describe the anisotropy of the attractive electron-electron coupling; the $\gamma_l^{(1F)}$ measure the anisotropy of the direction of the Fermi velocity, the density of states at the Fermi surface and the shape of the Fermi surface; the $\gamma_l^{(2)}$ characterize the anisotropy of $v^2(\underline{k}_F)$. The $h_l^{(i)}(t,\alpha)$ are weight functions, which characterize the weights of the various microscopic anisotropies contributing to the anisotropy of H_{c2}.

In (12) the coefficients c_l are introduced in order to obtain the asymptotic forms in the clean limit ($\alpha = 0$) for temperatures near T_c

$$h_l^{(1F)}(t,\alpha=0) \equiv h_l^{(1\varphi)}(t,\alpha=0) \underset{t \to 1}{=} (1-t)^{(1/2)-1}, \qquad (16)$$

$$h_l^{(2)}(t,\alpha=0) \underset{t \to 1}{=} \frac{1}{2} l\,(1-t)^{(1/2)-1}. \qquad (17)$$

For the lowest angular momentum values $l = 4$ and $l = 6$ the conditions (16) and (17) yield

$$c_{l=4} = \frac{3}{35} \cdot \frac{\lambda(5)}{\lambda(3)^2},$$

$$c_{l=6} = \frac{135}{2002} \cdot \frac{\lambda(7)}{\lambda(3)^3}.$$

The detailed (t,α)-dependence of the $h_l^{(i)}$ has to be determined numerically. Figure 3 presents plots of the weight functions for $l = 4$ and $l = 6$. (For convenience only the functions with $\alpha = 0, 0.1, 0.5$ and 1.5 are shown, more complete data will be given in /32/). A comparison of Fig.3a with Fig.3b shows, that $h_l^{(1F)}$ and $h_l^{(1\varphi)}$ are identical in the clean limit, and that $h_l^{(1\varphi)}$ decreases more rapidly with increasing impurity content than $h_l^{(1F)}$. From Fig.3c it is obvious that $h_l^{(2)}$ has a completely different temperature dependence even in the pure metal. Because of this result, values of the parameters $\gamma_{l=4}^{(2)}$ and $\gamma_{l=4}^{(1)} = \gamma_{l=4}^{(1F)} + \gamma_{l=4}^{(1\varphi)}$ may be extracted from measurements of the H_{c2} anisotropy in clean metals, whereas the further decomposition of $\gamma_{l=4}^{(1)}$ into $\gamma_{l=4}^{(1F)}$ and $\gamma_{l=4}^{(1\varphi)}$ (i.e. the decomposition into Fermi surface and electron-electron coupling contributions) may be achieved by an investigation of the impurity dependence of H_{c2} anisotropy /31, 32/. As may be seen from Fig. 3 d-f the weight functions with $l = 6$ display similar impurity dependences as those with $l = 4$. Therefore,

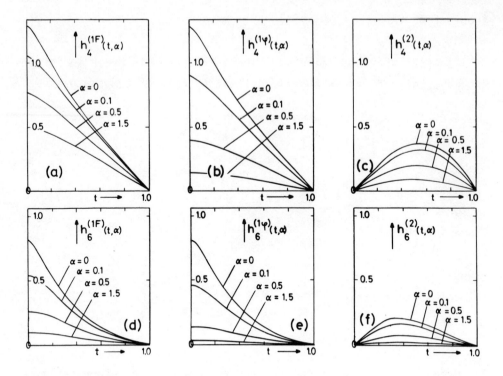

Fig.3: Weight functions $h_l^{(i)}(t,\alpha)$ describing the effects of microscopic anisotropies on $\Delta H_{c2}/\overline{H}_{c2}$

also the material parameters $\gamma_{l=6}^{(1F)}$, $\gamma_{l=6}^{(1\varphi)}$ and $\gamma_{l=6}^{(2)}$ may be deduced from experimental investigations of the temperature and impurity dependence of H_{c2} anisotropy.

5. Applications of the Nonlocal Theory to Nb and V*

The nonlocal theory of H_{c2} anisotropy has been applied to estimate the microscopic anisotropy parameters $\gamma_l^{(i)}$ of Nb /31/ and V /32/.

---

*The microscopic parameters $\gamma_l^{(i)}$ for Nb determined from recent H_{c2} anisotropy measurements of Seidl and Weber /38/ in the system Nb-N will be presented in /39/.

In the case of Nb the H_{c2} anisotropy data of Williamson /2/ and of Reed et al. /28/ obtained with samples of residual resistivity ratios $\Gamma = 750$ /2/ and 1600 /28/ were analysed in terms of the clean limit version of the nonlocal theory. Fitting expression (12) with $\alpha = 0$ to the low temperature ($t \leq 0.5$) data /2,28/ for a_4 and a_6 leads to /31/

$$\gamma_4^{(1)} \equiv \gamma_4^{(1F)} + \gamma_4^{(1\varphi)} = -0.248, \tag{18}$$

$$\gamma_6^{(1)} \equiv \gamma_6^{(1F)} + \gamma_6^{(1\varphi)} = 0.183. \tag{19}$$

The values of $\gamma_4^{(2)}$, $\gamma_6^{(2)}$ were estimated to be neglegibly small. As demonstrated in /31/ the parameter (18) describes well the observed temperature dependence of a_4 below $t = 0.5$, whereas the parameter (19) is capable to describe the temperature dependence of a_6 in the whole temperature range $0 < t \leq 1$.

For vanadium the $\gamma_1^{(i)}$ were estimated /32/ from H_{c2} anisotropy data by Williamson /2/ on a V sample with $\Gamma = 140$. According to /2/ this Γ-value corresponds to an impurity parameter $\alpha \approx 0.03$ or $\alpha \approx 0.1$, the precise value depending upon whether the electron mean free path is determined from transport phenomena or ultrasonic attenuation /2/. Due to this uncertainty the impurity parameter is treated in /32/ as an additional unknown parameter and the values of the material parameters and of α are determined by a least-squares fit of (12) to the experimental data. This procedure yields

$$\alpha = 0.093$$

$$\gamma_4^{(1\varphi)} = 0.10 \qquad \gamma_4^{(1F)} = -0.35 \qquad \gamma_4^{(2)} = -0.06 \tag{20}$$

$$\gamma_6^{(1\varphi)} = -0.01 \qquad \gamma_6^{(1F)} = 0.18 \qquad \gamma_6^{(2)} = -0.03 \tag{21}$$

Using these parameters good agreement between the experimental data and the theoretical description was obtained for reduced temperatures $t \leq 0.8$. However, as mentioned in /32/, the experimental data /2/ are not sufficient to determine the values of the material parameters $\gamma_1^{(i)}$ unambiguously: the evaluation of more reliable estimates needs further experimental investigations of the impurity dependence of H_{c2} anisotropy in vanadium.

6. Comparison of Results of the Quasilocal and Nonlocal Theories

According to Eq.(16) and ref. /32/ the asymptotic temperature dependence of the weight functions $h_1^{(i)}(t,\alpha)$ for $t \to 1$ is of the form

$$h_1^{(i)}(t,\alpha) \underset{t \to 1}{\propto} (1-t)^{\frac{1}{2}-1} \tag{22}$$

Therefore, the expansion coefficient $a_4(T,\tau)$ exhibits a linear $(1-t)$-dependence near T_c, whereas $a_6(T,\tau)$ shows a $(1-t)^2$-dependence. These asymptotic relationships agree with those obtained by Hohenberg and Werthamer /19/ for $l = 4$ and by Takanaka and Nagashima /6/ for $l = 6$. In the case of $l = 8$ Eq.(22) predicts an asymptotic $(1-t)^3$-dependence. This result seems to be in contradiction to /6/, where a $(1-t)^2$-dependence was derived for the leading term in a_8. However, a careful inspection of this discrepancy shows, that the $(1-t)^2$ term of a_8 in /6/ describes nonlinear effects of the microscopic anisotropies which, of course, are not present in the linearized nonlocal theory discussed above.

For temperatures below T_c the nonlocal theory /31,32/ as well as the quasilocal theories /6,19,29,30/ predict deviations of the a_l from their asymptotic $(1-t)^{(1/2)-1}$ forms. Hence, the range of applicability of the quasilocal theories may be estimated by comparing their predictions for the T-dependence, e.g., of $a_4(T,\tau \to \infty)$ with the result of the nonlocal theory /31/. According to the nonlocal theory the temperature dependence of $a_{l=4}$ in clean metals is characterized by two different weight functions $h_4^{(1F)}(t,\alpha=0) \equiv h_4^{(1\varphi)}(t,\alpha=0)$ and $\frac{1}{2} h_4^{(2)}(t,\alpha=0)$. By taking into account the first order nonlocal corrections, Hohenberg and Werthamer /19/ obtained one universal weight function for $l = 4$. Thus, according to (16), (17) their theory is valid only in the temperature range, where the splitting of $h_4^{(1F)}(t,0)$ and $\frac{1}{2} h_4^{(2)}(t,\alpha=0)$ is negligible, i.e. where the $h_l^{(i)}$ may be approximated by their asymptotic forms.

Takanaka and Nagashima /6/ considered first and second order nonlocal corrections. Linearization of their results with respect to the microscopic anisotropies shows, that in their theory the effects of microscopic anisotropies on the H_{c2} anisotropy are mediated by two different weight functions which correspond to those of the nonlocal theory and may, therefore, be denoted by $h_4^{(1F)}(t)_{TN}$ and $\frac{1}{2} h_4^{(2)}(t)_{TN}$. However, as indicated in Fig.4, the temperature dependence of these weight functions deviates appreciably from the t-dependence of the weight functions derived within the nonlocal theory. Thus, we have to conclude that even the quasilocal theory of Takanaka and Nagashima /6/ is valid only in a very narrow temperature range below T_c.

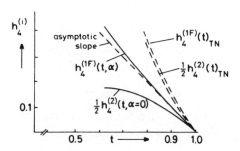

Fig.4: Weight functions $h_4^{(i)}(t,\alpha=0)$ of the nonlocal theory /31/ and corresponding functions $h_4^{(i)}(t)_{TN}$ resulting from the quasi-local theory /6/

7. Nonlocal Theory of Preferential Orientations of the Flux Line Lattice Relative to the Crystal Lattice

As shown schematically in Fig.1 the experimental results /10-12/ indicate preferential orientations of the FLL relative to the CL. In the following we describe briefly a recent nonlocal theory /40/ of this phenomenon, which is valid - at least - near H_{c1}, where the flux lines are well separated from each other.

The preferential orientations of the FLL indicate an orientation dependent interaction between flux lines. Moreover, the induction jump B_o observed in Nb /41,42/ shows that there exists an attractive interaction between flux lines in Nb /41,42/. According to the theory of Kramer /43/ and Leung /44/ every superconductor with κ slightly above $1/\sqrt{2}$ must have an attractive flux line interaction. Thus we may conclude, that flux lines in Nb or V affect each other by an anisotropic attractive interaction. This conclusion is confirmed by a recent theory of K.Fischer /45/, who proved for anisotropic superconductors with κ slightly above $1/\sqrt{2}$, that an attractive interaction energy between flux lines exists, which for large flux line distances r_{ij} has the asymptotic form

$$W_{ij} \sim -\exp(-r_{ij}/\xi(\underline{e}_{ij})) \tag{23}$$

$(\underline{r}_{ij} = r_{ij} \cdot \underline{e}_{ij})$. The decay length $\xi(\underline{e}_{ij})$ in (23) is identical with the anisotropic coherence length derived in /46/.

In order to evaluate the influence of the microscopic anisotropies on the orientation of the FLL we assume, that the structure of the FLL is known. If we restrict ourselves to nearest neighbor interactions in the FLL, the leading term in the orientation dependent part of the FLL energy for small anisotropies of the coherence length is given by

Microscopic Theory of H_{c2} Anisotropy

$$\Delta W_{FLL} \sim -(b/\bar{\xi}) \exp(-b/\bar{\xi}) \frac{1}{2} \sum_{ij} \frac{\Delta\xi(\underline{e}_{ij})}{\bar{\xi}} \qquad (24)$$

(b: nearest neighbor distance). $\Delta\xi(\underline{e}_{ij})/\bar{\xi}$ denotes the relative anisotropy of the coherence length. For impurity-free superconductors, $\Delta\xi/\bar{\xi}$ is given in linear order of the microscopic anisotropies by

$$\frac{\Delta\xi(\underline{e})}{\bar{\xi}} = \sum_{l=4,6,...} H_l(\underline{e})\{g_l^{(1F)}(t)\gamma_l^{(1F)} + g_l^{(1\varphi)}(t)\gamma_l^{(1\varphi)} + g_l^{(2)}(t)\gamma_l^{(2)}\} \qquad (25)$$

where the $\gamma_l^{(i)}$ are identical to those entering the expression (12) for the relative H_{c2} anisotropy. The $g_l^{(i)}(t)$ denote the weights by which the various $\gamma_l^{(i)}$ contribute to $\Delta\xi/\bar{\xi}$ and thus to the orientation dependent part of the FLL energy. The results of numerical calculations for these functions are presented in Fig.5 for the angular momenta l = 4 and l = 6. Obviously the $g_l^{(i)}$ are completely different from the $h_l^{(i)}$. In particular, the $h_l^{(i)}$ are always positive, whereas the $g_l^{(1\varphi)}$ change their signs at intermediate temperatures. (Because of this result metals with similar H_{c2} anisotropy may show different

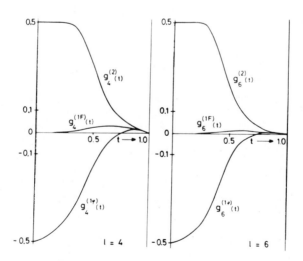

Fig.5: Weight functions $g_l^{(i)}(t)$ describing the effects of microscopic anisotropies on $\Delta\xi/\bar{\xi}$

preferential orientations of the FLL relative to the CL at low temperatures.) Nevertheless, for $t \to 1$ the asymptotic proportionalities $g_1^{(i)} \sim (1-t)^{(1/2)-1}$ hold.

Using the $\gamma_1^{(i)}$ values of (20) and (21) for V and assuming a triangular FLL, Eq.(24) yields for magnetic fields parallel to the <111>, <110>, or <112> crystal axes, that at low temperatures one of the basis vectors of the FLL should be parallel to the <1$\bar{1}$0> direction. This theoretical result agrees with experimental observations on V single crystals /47/. In addition, if for magnetic fields parallel to the <001> crystal axis a square FLL is realized, Eq.(24) predicts that the basis vectors of the FLL are parallel to the crystallographic <110> directions.

An application of this nonlocal theory of preferential orientations to Nb is given in /39/, where the complete set of microscopic anisotropy parameters $\gamma_1^{(i)}$ deduced from recent H_{c2} anisotropy data of Seidl and Weber /38/ will be presented.

The author gratefully acknowledges stimulating remarks of Prof.Dr.A. Seeger and his encouraging interest in this problem.

References

/1/ D.R.Tilley, G.J. van Gurp, C.W.Berghout, Phys.Lett. <u>12</u>, 305 (1964)
/2/ S.J.Williamson, Phys.Rev. <u>B2</u>, 3545 (1970)
/3/ G.Kostorz, L.L.Isaacs, R.L.Panosh, C.C.Koch, Phys.Rev.Lett. <u>27</u>, 304 (1971)
/4/ C.E.Gough, Solid State Commun. <u>6</u>, 215 (1968)
/5/ I.Williams, A.M.Court, Solid State Commun. <u>7</u>, 169 (1969)
/6/ K.Takanaka, T.Nagashima, Progr.Theor.Phys. <u>43</u>, 18 (1970)
/7/ H.Kiessig, U.Essmann, H.Teichler, W.Wiethaup, Proc. 14th Internat. Conf. Low Temp. Phys., Otaniemi, Finland 1975, vol.2, p. 317

/8/ K.Takanaka, A.Hubert, Proc. 14th Internat. Conf. Low Temp. Phys., Otaniemi, Finland 1975, vol.2, p. 309
/9/ H.Träuble, U.Essmann, phys.stat.sol. 18, 813 (1966)
/10/ J.Schelten, G.Lippmann, H.Ullmaier, J.Low Temp.Phys. 14, 213 (1974)
/11/ U.Essmann, Physica 55, 83 (1971)
/12/ B.Obst, phys.stat.sol.(b) 45, 467 (1971)
/13/ L.P.Gor'kov, T.K.Melik-Barkhudarov, Soviet Phys. JETP 18, 1031 (1964)
/14/ C.Caroli, P.G. de Gennes, J.Matricon, Physik kondens. Materie 1, 176 (1963)
/15/ H.Teichler, Phil.Mag. 30, 1209 (1974)
/16/ L.P.Gor'kov, Soviet Phys. JETP 7, 505 (1958)
/17/ H.Teichler, Phil.Mag. 31, 775 (1975)
/18/ K.Takanaka, paper R-3, this volume, p. 93
/19/ P.C.Hohenberg, N.R.Werthamer, Phys.Rev. 153, 493 (1967)
/20/ W.E.Lawrence, S.Doniach, Proc. 12th Internat. Conf. Low Temp. Phys., Kyoto 1970, p. 361
/21/ L.N.Bulayevskii, Soviet Phys. JETP 38, 634 (1974)
/22/ R.A.Klemm, A.Luther, M.R.Beasley, Phys.Rev. B12, 877 (1975)
/23/ K.Takanaka, H.Ebisawa, Progr.Theor.Phys. 47, 1781 (1972)
/24/ K.Takanaka, phys.stat.sol.(b) 68, 623 (1975)
/25/ H.Teichler, phys.stat.sol.(b) 72, 211 (1975)
/26/ M.R.Skokan, R.C.Morris, W.G.Moulton, Phys.Rev. B13, 1077 (1976)
/27/ F.C. von der Lage, H.A.Bethe, Phys.Rev. 71, 612 (1947)
/28/ W.A.Reed, E.Fawcett, P.P.M.Meincke, P.C.Hohenberg, N.R.Werthamer Proc. 10th Internat. Conf. Low Temp. Phys., Moscow 1966, vol.2A, p. 368
/29/ K.D.Harms, Z.Naturforsch. 25a, 1161 (1970)
/30/ T.Nagashima, Progr.Theor.Phys. 47, 37 (1972)
/31/ H.Teichler, phys.stat.sol.(b) 69, 501 (1975)
/32/ H.W.Pohl, H.Teichler, phys.stat.sol.(b) 75, 205 (1976)
/33/ G.Eilenberger, Z.Physik 214, 195 (1968)

/34/ O.Betbeder-Matibet, P.Nozieres, Ann.of Phys. $\underline{51}$, 392 (1969)
/35/ D.Markowitz, L.P.Kadanoff, Phys.Rev. $\underline{131}$, 563 (1963)
/36/ P.Hohenberg, Soviet Phys. JETP $\underline{18}$, 834 (1964)
/37/ J.R.Clem, Phys.Rev. $\underline{148}$, 392 (1966)
/38/ E.Seidl, H.W.Weber, paper C-2, this volume, p.57
/39/ H.Teichler, paper C-3, this volume, p.65
/40/ K.Fischer, H.Teichler, submitted to Phys.Lett.
/41/ U.Kumpf, phys.stat.sol.(b) $\underline{44}$, 829 (1971)
/42/ J.Auer, H.Ullmaier, Phys.Rev. $\underline{B7}$, 136 (1973)
/43/ L.Kramer, Z.Physik $\underline{258}$, 367 (1973)
/44/ M.C.Leung, J.Low Temp.Phys. $\underline{12}$, 215 (1973)
/45/ K.Fischer, Thesis, Stuttgart 1975
/46/ H.Teichler, Phil.Mag. $\underline{31}$, 789 (1975)
/47/ M.Roger, R.Kahn, J.M.Delrieu, Phys.Lett. $\underline{50A}$, 291 (1974)

R-2   EXPERIMENTAL INVESTIGATIONS ON THE ANISOTROPY OF THE UPPER CRITICAL FIELD IN TYPE II SUPERCONDUCTORS

T.Ohtsuka

Faculty of Science, Department of Physics

Tohoku University, Sendai 980, Japan

Abstract

The present status of the experimental investigation on the upper critical field H_{c2} in cubic materials is reviewed. Although a complete quantitative correspondence with theories based on nonlocality is not possible because of unknown parameters involving Fermi surface anisotropy and the anisotropic attractive interaction, experiments up to date are in general agreement with the predicted dependence of anisotropy on temperature and impurity scattering. Investigation on the anisotropic temperature dependence of H_{c2} at low temperatures, the average critical field at absolute zero and anisotropy of the lower critical field is also briefly reviewed.

1. Introduction

Among a variety of anisotropic phenomena observed in superconductors, this review is concerned mainly with the anisotropy of one of the fundamental parameters of type II superconductors, the upper critical field H_{c2}. The report will be confined to investigations on materials with cubic symmetry. Crystal symmetry is of special importance here, because no H_{c2} anisotropy is expected for cubic materials within the framework of the local Ginzburg-Landau theory.

The first clear observation of H_{c2} anisotropy in cubic materials was reported by Tilley et al. /1/ through magnetization measurements on a Nb single crystal sphere. The analysis of the results was based on the anisotropic Ginzburg-Landau (GL) equation derived by Ginzburg /2/ and its extension by Tilley /3/. Subsequent to the work of Ginzburg, Gor'kov and Melik-Barkhudarov /4/ and also Caroli et al. /5/ provided a microscopic basis for the anisotropic GL equation. In both the phenomenological and the microscopic theory anisotropy appears through the replacement of the effective mass by an effective mass tensor m_{ij}. The microscopic derivations, however, show that $1/m_{ij}$ is a second rank tensor proportional to the dc electric conductivity tensor /5/. This implies that no anisotropy is expected in materials with cubic symmetry in the local GL limit.

This fact was first pointed out by Hohenberg and Werthamer /6/ (hereafter HW), who also showed that Fermi surface anisotropy can give rise to H_{c2} anisotropy by considering nonlocal corrections to the local GL equation derived by Gor'kov /7/. The theory thus predicts a diminution of anisotropy as the local limit is approached with an increase in temperature and impurity scattering. The main features of the HW theory were found to be in accord with early experiments by Reed et al /8/.

The fact that nonlocality plays a primary role in determining anisotropy makes investigations on cubic materials interesting, since they present a sensitive probe of checking the validity of nonlocal microscopic theories. Moreover, there exists the possibility that interesting parameters related to the Fermi surface may be extracted by a proper analysis of experiments, if the validity of theory can be established. There already exists a fair amount of experimental work on H_{c2} anisotropy which, in general, supports theories based on nonlocality. A complete quantitative check of the theory is, however, difficult at present, because the anisotropy parameters include a factor depending on the variation of the Fermi velocity over the Fermi surface and the anisotropic interaction which is unknown.

It is noted, however, that there is one quantity, namely the enhancement of the value of the reduced spatially averaged upper critical field at absolute zero due to anisotropic effects, which appears to be in good quantitative agreement with Fermi surface calculations. This will be discussed later together with anisotropic properties of other quantities intimately related to the upper critical field, after a review of investigations made on H_{c2} anisotropy has been presented.

2. Experimental - Methods and Analysis

Except for the need of using single crystals and methods allowing a variation of the field direction relative to the crystallographic axes, conventional techniques for measuring H_{c2} may be employed to determine its anisotropy. Magnetization, resistance, magnetothermal and torque methods have been used.

Experiments are usually performed by rotating the magnet or the sample so that the field direction lies in a crystal plane containing the three principal axes <100>, <110> and <111>, and determining H_{c2} for successive field orientations at a fixed temperature. In order to deduce useful parameters from the measured anisotropy curves, they are analysed by fitting the data to an expansion in cubic harmonics $H_n(\alpha,\beta,\gamma)$,

$$H_{c2} = \sum_n A_n H_n(\alpha,\beta,\gamma) \qquad (1)$$

Here α, β, γ are direction cosines of the field direction with respect to the cubic axes.

Except for normalization factors the cubic harmonics initially used by Reed et al. in their analysis are essentially the same as the expansion introduced by Von der Lage and Bethe /12/. The explicit form of Eq.(1) up to $n=4$ is

$$\begin{aligned}
H_1 &= 1 \\
H_2 &= 5.728\,(\alpha^4+\beta^4+\gamma^4-0.6) \\
H_3 &= 147.2\,(\alpha^2\beta^2\gamma^2+0.00794\,H_2-0.00952) \\
H_4 &= 96.22\,(\alpha^8+\beta^8+\gamma^8-0.0380\,H_3-0.2565\,H_2-0.3333)
\end{aligned} \qquad (2)$$

In (2) the numerical factors are adjusted so that the squared modulus of H_n integrated over all directions is normalized to 4π.

The coefficient A_1 of the $n=1$ term corresponds to the directionally averaged upper critical field $\overline{H_{c2}}$. As will be discussed later, the limits of experimental accuracy infer that an expansion up to $n=4$ is sufficient to fit the measured anisotropy curves, even for clean samples at low reduced temperatures.

Parameters of physical interest are contained in the coefficient A_n. They are, in general, complicated functions of microscopic anisotropies related to the Fermi surface and the superconducting interaction, the temperature and impurity. There are two approaches in calculating A_n, one which assumes small nonlocality and one which retains full nonlocality but assumes anisotropy to be sufficiently small. Calculations based on small nonlocality or quasilocal conditions were first made by HW and extended by Takanaka and Nagashima /9,10/ to include higher nonlocal terms, and by Teichler

/11/ who included the anisotropic interaction in addition to Fermi surface anisotropy. The validity of theory is restricted to the nearly local regime where an expansion parameter ε, given by

$$\varepsilon = \frac{2e\, H_0(t)\, v_F^2\, \hbar}{c(2\pi k_B T_c)^2} (t+\alpha)^{-2} \tag{3}$$

is sufficiently small compared to unity. Here H_0 is the critical field in the local limit, which is approximately the same as the average critical field in the case of small nonlocality, v_F the Fermi velocity and α the impurity parameter defined by

$$\alpha = \frac{\hbar}{2\pi k_B T_c \tau} = 0.882 \frac{\xi_0}{\ell} \tag{4}$$

where τ is the impurity scattering time, ξ_0 the coherence length and ℓ the mean free path.

On the other hand, the calculations based on small anisotropy are valid in the entire nonlocal regime. Calculations are, however, more complicated and only clean limit results have been given. HW have shown that a relatively simplified expression can be given for the average reduced field at absolute zero $\overline{h^*}(0)$ /30/

$$\overline{h^*}(0) = \frac{\overline{H_{c2}}(0)}{-\left(\frac{dH_{c2}}{dt}\right)_{t=1}} \tag{5}$$

More recently, Teichler /11/ has extended calculations to finite temperatures.

The results obtained in both approaches can be casted into a form,

$$\frac{A_n}{A_1} = \sum_i \{B_n^i\, G_n^i(t,\alpha) + \text{(contributions from higher order terms)}\} \tag{6}$$

Here i denotes the origin of anisotropy. Teichler showed, that for the leading term contributions from different microscopic sources of anisotropy can be expressed as a sum of parameters factorized into a material parameter B_n^i and a universal function $G_n^i(t,\alpha)$, which depends only on temperature and impurity. As the material parameter which is a complicated function of Fermi velocity and interaction anisotropy is not known apriori, comparison with theory can be made only through an analysis of the universal function $G_n^i(t,\alpha)$. This will be discussed in the following sections.

3. Experiments on H_{c2} Anisotropy

3.1 Experiments on Pure Samples

Both Nb and V, the only two known intrinsic type II superconductors with cubic symmetry, have been investigated. In particular, Nb has been investigated in detail by several authors. The residual resistance ratios (RRR) of the Nb samples investigated range from RRR = 100 to 1600.

The first cubic harmonic analysis of the anisotropy curves was reported by Reed et al. /8/. The niobium sample they used had an RRR = 1600, and they showed, that the anisotropy could be described fairly well using cubic harmonics up to n = 3. Subsequently, Gough /13/ working with three samples with RRR = 650, 920 and 1450, respectively, found agreement with the results of Reed et al at 1.2 K. Similar results were obtained by Williams and Court /14/ on three Nb samples originating from a parent crystal having RRR = 250. They moted that although the three samples differed greatly in flux pinning behavior, the H_{c2} anisotropy was essentially the same as it should be.

Except for the study by Reed et al. the anisotropy measurements in the works quoted above were restricted to one or two fixed temperatures. Using a sample with RRR = 100, Farrell et al. /15/ made a detailed study of the H_{c2} anisotropy covering a temperature range from 1.2 K to the transition temperature. They find that above $t \gtrsim 0.5$ a four term (n = 4) cubic harmonic fit reproduces the data to experimental accuracy. For $t < 0.5$ they state that small systematic deviations from the n = 4 fit arise, but reliable values of A_5 could not be obtained. Similar observations were made on a Nb sample with RRR = 750 by Williamson /16/, who extended measurements down to 0.055 K (t = 0.006). Fig.1 shows an example of a four parameter fit at t = 0.05 made by Williamson. The deviation appears to be largest near the <001> direction. Williamson notes that a five term fit results in little improvement only. Yamamoto et al. /17/ have also measured a sample with RRR = 600 from 1.2 K to near T_c and find overall agreement with the results obtained by Reed et al., Farrell et al. and Williamson over the temperature range covered, especially for $A_1(t)$ and to a lesser degree for $A_2(t)$. For $A_3(t)$ the values obtained by Farrell et al. lie distinctly below those of other authors, the deviation increasing with decreasing temperature. This is probably due to the purity of the sample (RRR = 100) used as already noted by Farrell et al. The results infer that, at least up to the third harmonic coefficient, samples with RRR above several

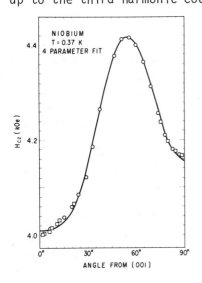

Fig.1:
Anisotropy curve of H_{c2} in the (110) plane for Nb at t = 0.05. The solid line represents a fit by four cubic harmonics (after Williamson /16/).

hundred can be considered to represent the clean limit.

As for the temperature dependence of the coefficients A_n, the quasilocal theory predicts $A_2 \propto (1-t)^2$ and $A_3 \propto (1-t)^3$. According to Takanaka and Nagashima, $A_4 \propto A_3 \propto (1-t)^3$. For pure Nb the expansion parameter ε (Eq.(3)) which defines the validity regime for the quasilocal theory becomes smaller than unity only for reduced temperatures $t > 0.7$. Thus, one has to work quite closely to the transition temperature, where errors become large due to the smallness of anisotropy. Experimental data in this region are very few and a strict confirmation of theory cannot be said to have been established. However, experiments to date appear to support qualitatively the predictions for the temperature dependence of A_2 and A_3. In fact, all experiments indicate that the power law $A_n \propto (1-t)^x$ is obeyed down to the lowest temperatures investigated so far with $x \simeq 2$ and $x \simeq 3$ for $n = 2$ and $n = 3$, respectively. The predicted temperature insensitivity of the ratio A_4/A_3 also appears to be supported down to low temperatures /15,16/.

The fact that the quasilocal limit behavior appears to extend down to low temperature regions has, to a large extent, been clarified by Teichler /11/ who, based on the small anisotropy approach, calculated the temperature dependence of A_2 and A_3 for the clean limit valid in the nonlocal regime. Although analytical forms are not given, numerically calculated curves show that the pertinant temperature dependence function has features similar to the asymtotic expansions valid for the nearly local limit. Fitting the low temperature data of Reed et al. and Williamson in the region $t \lesssim 0.5$, good agreement is obtained for A_3 in the entire temperature region covered, whereas for A_2 the experimental points deviate upwards from the theoretical curve at higher temperatures.

In comparison with Nb there are only relatively few investigations on V. Moreover, it is difficult to obtain samples with the same degree of purity as with Nb, which casts some doubts to whether measurements really reflect the clean limit. For instance, the sample measured by Williams and Court in their early work is quoted to have an impurity parameter $\alpha = 2.04$. The sample used by Williamson in his detailed measurements had RRR = 140. He observes, however, that the values of A_2/A_1 for all temperatures are similar to those obtained for cleaner Nb samples (RRR = 750 and 1600), which infers that purer V has a larger relative anisotropy than pure Nb. For A_3, the values lie distinctly below those of Nb indicating the effect of impurity scattering. Beside of this discrepancy the overall features of the anisotropy for V and Nb are similar. Anisotropy of H_{c2} for less pure V samples (RRR = 87) has been determined by Hembach et al. through torque studies /18/.

The H_{c2} anisotropy of two compounds, V_3Ge and V_3Si, has been measured by Reed et al., who find that the sign of A_2 is opposite to that of Nb. Although compounds which are ideally stoichiometric may represent the pure limit case, it is doubtful whether this can be realized in practice.

3.2 Impurity Effect

As nonlocality effects are reduced with impurity as well as with temperature, systematic studies of the impurity effect on the H_{c2} anisotropy should also provide important information on the mechanism leading to anisotropy. The dependence of anisotropy on the impurity parameter α defined by Eq.(3) was first studied by Reed et al. who compared the coefficient A_2 for pure Nb ($\alpha \ll 1$) with that of $Nb_{99}Ta_1$ ($\alpha \sim 0.2$) and $Nb_{50}Ta_{50}$ ($\alpha \sim 4$). In qualitative accordance with the prediction of theory A_2 was found to decrease with increasing α. As mentioned earlier, Farrell et al. /15/ noted, that impurity not only reduces the overall anisotropy but affects the higher harmonics more drastically.

A detailed study of the impurity effect was first made by Yamamoto et al. /17/ on Nb-Ta and Nb-Mo alloys, who also observed similar trends. They found, that the coefficient of the second harmonic, A_2, decreased rapidly upon increasing the impurity parameter up to about $\alpha \sim 1$ and tended to flatten off for larger values of α. As for A_3, the reduction with increasing α was more drastic, A_3 becoming practically negligible for samples with $\alpha > 1$. An analysis was made on the dependence of A_2 on α considering only the leading term in Eq.(6)

$$\frac{A_2}{A_1} = B_2 G_2(\alpha, t) \tag{7}$$

and assuming that the material parameter B_2 did not change appreciably with the addition of Ta or Mo. By normalizing to pure Nb they found, that the α-dependence of the universal function $G_2(\alpha, t)$ was in reasonable agreement with quasilocal theory calculations at $t = 0.7$. For lower reduced temperature, the experimental results for $\alpha > 1$ lied appreciably above the theoretical curves. In the light of recent measurements made on Pb-In alloys to be discussed shortly, this finding is now considered as reflecting the fact that the normalization of the experimental and theoretical results was made with pure Nb, where the quasilocal theory is invalid, the expansion parameter ε being larger than unity for $t < 0.7$.

In addition to the problem discussed above, contributions from possible anisotropic interactions should become more pronounced in the low α region. For Nb-alloys there exists another problem in analyzing the results, namely the relatively large change of the transition temperature T_c with alloying, which is not taken into account by previous theories. This problem, however, has been discussed recently by Nagashima and Fukuyama /19/, who showed that the expression for H_{c2} is unchanged when T_c and the Fermi velocity are replaced by those of the alloys. Due to various problems involved and in view of a comparison with theory it is preferable, however, to perform detailed measurements on systems, in which quasilocal conditions are satisfied and the change in T_c is relatively small for a wide range of α-values. Pb-In is such a system and a detailed analysis has been made by Ohta /20/ recently.

The Pb-In system is known to be an extrinsic type II superconductor for In contents above about 1 at%. A primary cubic (fcc) phase exists up to about 65 at% In allowing a relatively wide variation of the impurity parameter α. Through detailed investigations on Pb-In polycrystals, Farrell et al. /21/ have shown that α varies from $\alpha \sim 2$ to $\alpha \sim 19$ ($Pb_{50}In_{50}$) which is to be compared with the largest value for Nb-Ta of about $\alpha \sim 4$ ($Nb_{50}Ta_{50}$). Moreover, for Nb-Ta T_c varies by 33% from $\alpha \ll 1$ to $\alpha \sim 4$, whereas the corresponding change in Pb-In is only about 0.8%.

The H_{c2} anisotropy was measured by Ohta on six samples with the In-content ranging from 1 at% to 20 at% covering the impurity parameter range $2 < \alpha < 14$. Preliminary results have been published /22/ and showed that the second harmonic coefficient A_2 became virtually independent of impurity content above about 2 at% In or for $\alpha > 3$. In contrast to the statement made there, the fact that A_2 remains constant in the quasilocal regime is precisely what HW predict as it was found that the average critical field, $\overline{H_{c2}} = A_1$, varied linearly with impurity parameter. Therefore, if the variation of the material parameter B_2 with alloying is assumed to remain sufficiently small, $G_2 \propto 1/\alpha$, which is the conclusion of the quasi-local theory. The experimental results for A_2/A_1 versus α at $t = 0.2$ are shown in Fig.2, where the solid curve represents numerical calculations made by Nagashima /22/ for G_2. As the values of T_c and $\overline{H_{c2}}$ agreed very well with the measurements made by Farrell et al. /21/, the values of the impurity parameter derived by them were used. It is seen that the agreement between experiment and theory

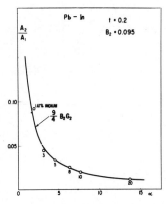

Fig.2:
The impurity parameter dependence of A_2/A_1 for Pb-In alloys at $t = 0.2$. The factor (9/4) is included to make correspondence to Hohenberg and Werthamer's expression for B_2. The solid curve represents numerical calculations for the universal function G by T. Nagashima scaled to give $B_2 = 0.095$ which is assumed constant.

is very good establishing the validity of theory at least in the quasilocal regime and within the assumptions made. It is noted that the expansion parameter for the cleanest sample ($\alpha \simeq 2$) is $\varepsilon \simeq 0.9$. A more rapid reduction rate of A_3 with impurity was also observed but has not been analysed.

Similar results were obtained for Pb-Tl alloys by Roger et al. /24/ who measured the H_{c2} anisotropy on samples with Tl contents of 1.6, 3, 10 and 20 at%, respectively. They showed that at 1.2 K the value of A_2 was virtually independent of Tl content but that A_3 diminished rapidly becoming negligible for a 20 at% Tl sample.

As the GL parameter κ_1 decreases somewhat with increasing temperature, there exist samples which undergo a transition from type I to type II superconductivity as the temperature is reduced. An interesting observation in connection with this transition was reported by Kiessig et al. /25/ recently, who observed that for a Pb-1.8 at% Tl sample, which undergoes a type I to II transition at $T^* = 4.6$ K, extrapolation of the reduced coefficient A_2/A_1 from $T < 4.2$ K indicates that anisotropy does not vanish at the transition temperature T^*. This implies that if T^* is defined by $H_{c2}(\alpha,\beta,\gamma;T^*) = H_c(T^*)$, there exists an anisotropic effect in T^*. In other words, the transition temperature depends on the field direction with respect to the crystal axis.

As for the temperature dependence of the reduced coefficients A_2/A_1 and A_3/A_1, there appears to be no difference within the clean limit results in regions where a comparison can be made. In other words, temperature and impurity act independently in reducing non-locality.

4. Anisotropy Effect in Quantities Related to H_{c2}

4.1 Temperature Dependence of H_{c2} at Low Temperatures

The temperature dependence of H_{c2} at temperatures close to absolute zero can usually be described by a quadratic relation $H_{c2}(t) = H_{c2}(0)(1 - at^2)$. Not only $H_{c2}(0)$ but also the coefficient a is anisotropic, the microscopic origin being related to the anisotropy of H_{c2}. Unfortunately measurements on the orientation dependence of a are lacking and, to the author's knowledge, the only measurements which extend down to very low temperatures are those by Williamson and Valby /26/, who measured the temperature dependence of H_{c2} for the three crystallographic directions <001>, <110> and <111> on Nb(RRR = 750) and V (RRR = 140) down to 0.055 K. In the temperature range $0 < t < 0.2$, they found that for Nb the experimental results for all three orientations could be fitted well to Gor'kov's clean limit relation /7/ $H_{c2}(t) = H_{c2}(0)(1 + \eta t^2 \ln t)$ instead of the quadratic relation. Similar results were obtained for V, but the fit to the $t^2 \ln t$ relation was not as good. For Nb they give $\eta = 1.69$, 1.39 and 1.30 for the <111>, <110> and <001> directions, respectively. Assuming that the anisotropy of η in V is the same as with Nb, Williamson /16/ has shown that the results for V could be fitted to a relation, where $\ln t$ is replaced by $\ln(t + \hbar/k_B \tau T_c)$ with $\hbar/k_B \tau \simeq 1$ K. The values of η obtained are about twice the value calculated by Gor'kov for the isotropic clean limit, $\eta = 0.65$.

The anisotropy of η has been discussed by Takanaka /27/ based on the small anisotropy approach. Recently he has pointed out that the discrepancy of η with Gor'kov's value may be removed to a large extent by fitting the data to a more accurate relation for $t \ll 1$,

$$H_{c2}(t)/H_{c2}(0) = 1 + \eta t^2 \ln t - \alpha t^2. \tag{8}$$

The t^2 term was neglected in Williamson's analysis, because the value of the coefficient was unknown. Based on the expression given by Maki and Tsuzuki /28/, Takanaka /29/ has calculated α to be $\alpha = 0.854$. Thus within a large portion of the temperature range where the data was fitted, the t^2 term is of the same order as the $t^2 \ln t$ term and cannot be neglected. Further discussion of this quantity must therefore await reanalysis of experimental results.

4.2 The Average Reduced Field $\overline{h^*(t)}$

A notable characteristic of the average reduced field $\overline{h^*(t)}$ defined by Eq.(5) is that, when normalized to the slope at $t = 1$, the measured values for pure Nb and V lie considerably above the values calculated by Helfand and Werthamer /30/ for an isotropic superconductor over the entire temperature range. Similar observations were made for the GL parameter κ_1. In fact, it had been known for some time that the value of κ_1 at absolute zero for Nb and V was considerably larger than the theoretical isotropic clean limit value $\kappa_1(0)/\kappa = 1.25$. This enhancement was attributed to anisotropy by HW who, based on the small anisotropy approach, obtained expressions for $\overline{h^*(0)}$ and showed that Fermi surface anisotropy may lead to an enhancement over the isotropic clean limit value $\overline{h^*(0)} = 0.727$ /30/.

Values of $\overline{h^*(0)}$ have been determined for pure Nb and V and Nb-Ta alloys. For pure Nb Ohtsuka and Kimura /31/ give $\overline{h^*(0)} = 0.85$ from measurements on a polycrystal and Williamson /16/ and Yamamoto et al. /17/ give $\overline{h^*(0)} = 0.96 \pm 0.03$ and 0.89, respectively, from measurements on single crystals. The rather large discrepancy among the reported values is possibly due to errors involved in the determination of the slope $(dH_{c2}/dt)_{t=1}$. Williamson /32/ notes that below $t \simeq 0.9$ (or $H_{c2} > 400$ G), H_{c2} in all directions deviates upward from a linear dependence on temperature. The value determined by Williamson /16/, who also obtained $\overline{h*(0)} = 0.93$ for V, agrees very well with $\overline{h*(0)} = 0.99$ obtained by Mattheiss /33/ for Nb from band structure calculations using HW's expression. This is probably the only result which shows quantitative agreement with the nonlocal theory /34/.

The effect of impurity on $\overline{h*(0)}$ has been reported by Ohtsuka and Kimura /31/ who determined $\overline{h*(0)}$ for Nb-Ta alloys with Ta contents ranging from 2 to 30 at%. After a rapid initial reduction from the pure Nb value, $\overline{h*(0)}$ decreases almost linearly up to $\alpha \sim 3.5$ (30 at% Ta), where the observed value $\overline{h*(0)} = 0.74$ is still appreciably larger than the calculated dirty limit isotropic value $\overline{h*(0)} = 0.69$. Unfortunately, at present no calculation for the impurity effect on $\overline{h*(0)}$ is available, which may be compared with these observations.

4.3 The Lower Critical Field H_{c1}

As the thermodynamic critical field H_c, which is proportional to the area under the magnetization curve, should not depend on field orientation, one naively expects that the anisotropy of H_{c2} leads to an anisotropy of H_{c1} with opposite sign. Recent calculations by Takanaka and Hubert /35/ show, that this is qualitatively correct and that the anisotropy of H_{c1} should be smaller than that of H_{c2}. They quote that the calculations are in good agreement with preliminary experiments.

Measurements on H_{c1} anisotropy are lacking. Whereas sample shape does not affect the measurements of H_{c2}, it affects H_{c1} considerably and the samples must be shaped so that the demagnetization coefficients are equal for all field directions in which measurements are made.

Recently Ohta /20/ made measurements of H_{c1} on pure Nb and V single crystal spheres, where in contrast to expectations a complicated anisotropy curve was observed when the field direction was rotated in the (110) plane. An example of an anisotropy curve for V is shown in Fig.3. A similar curve was also obtained for Nb. The relative anisotropy was about 1.6% for Nb at $t \simeq 0.2$ and 2.8% for V at $t \simeq 0.25$, much smaller than for H_{c2}. The anisotropy curves obtained cannot be fitted to an expansion with cubic harmonics at least up to n = 4.

A careful inspection of the virgin magnetization curves near H_{c1} showed that, whereas no anisotropy existed in the Meissner state, as expected, the magnetization curves above and near H_{c1} varied linearly with magnetic field with identical slopes $\Delta M/\Delta H = 3/4\pi$ for all field directions. This indicates the fact, that the H_{c1} measured is not the transition field to the vortex state but to the so-called

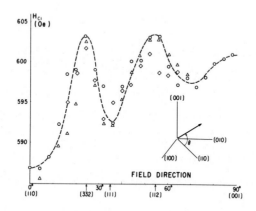

Fig.3: Anisotropy of the lower critical field H_{c1} of vanadium in the (110) plane at T = 1.4 K. H_{c1} was always determined from the virgin magnetization curve. Different symbols denote measurements made on different runs.

intermediate mixed state in the spherical samples. At present, there is no theory of anisotropic critical fields pertaining to this type of first order transition. However, since the transition field, which we denote by H^*_{c1}, is determined by the balance of free energies, one expects, that an anisotropy in the surface energy between the vortex and Meissner states in the intermediate mixed state region should be reflected by H^*_{c1}. In this respect, it is interesting to note, that the anisotropy of the surface energy determined by Obst /36/ for a Pb-Tl sample with $\kappa = 0.72$ is qualitatively similar to the anisotropy of H^*_{c1} for Nb as shown in Fig.4 except for a minimum between the <112> and <001> directions. The patterns are displaced by 90°, because the surface direction is defined by its normal, whereas the surface energy reflected in H^*_{c1} corresponds to field directions parallel to the specific surface. The upward trend of H^*_{c1} near <001> may be due to anisotropy related to the H_{c2} anisotropy superposed on the observed patterns.

Experimentally, more work is needed to establish this interpretation unambiguously. In particular, measurements on higher κ materials, where no intermediate mixed state exists, are necessary. Although the fact that Nb, which showed a larger hysteresis near H_{c1}

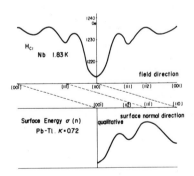

Fig.4: Comparison of the anisotropy of the lower critical field H_{c1} for Nb at $T = 1.83$ K with the anisotropy of the surface energy σ for Pb-Tl as determined by Obst /36/.

than V, showed smaller anisotropy, and that the slopes of the magnetization curves above and near H_{c1}^{*} showed no anisotropy, infers that flux pinning is not the origin of anisotropy, careful elimination of all extrinsic sources of anisotropy should be made.

5. Summary

A review has been given on the experimental investigation of the anisotropy of H_{c2} and related quantities in cubic type II superconductors. H_{c2} anisotropy in cubic materials is of interest, as it is one of the few if not the only phenomenon which depends solely on nonlocal properties of the order parameter, a property of vital importance for our understanding of superconductivity. It can be stated, that the essential correctness of H_{c2} anisotropy theories based on nonlocality has been established through the analysis of the temperature and impurity scattering dependence of the anisotropy parameters. Present experiments, however, cannot distinguish between the microscopic origins of anisotropy, and the material parameters are subject to future analysis.

As for the anisotropic temperature dependence of H_{c2} near absolute zero and the value of $\overline{h^{*}(0)}$, more work is necessary to evaluate the physical significance of the observed parameters. The role of impurity in this nonlocal regime is not clear both experimentally and theoretically. There is also much to be clarified in the anisotropy of the lower critical field H_{c1}. Although some work has appeared on the anisotropy of magnetization /13,14,18/, it has not been included in the review, because it is felt that more systematic work is necessary before a thorough analysis can be made. Future work on all these quantities should hopefully provide a coherent picture as to how microscopic anisotropy is reflected in macroscopic anisotropy related to the critical field.

The author wishes to thank Drs. K.Takanaka, T.Nagashima, N.Ohta and Professors H.Fukuyama and T.Tsuzuki for discussions in preparing this manuscript and Professor S.J.Williamson for his kind permission in allowing me to use his figure.

References

/1/ D.R.Tilley, G.J. van Gurp, and C.W.Berghout, Phys.Lett. 12, 305 (1964)
/2/ V.L.Ginzburg, Zh.Eksp.i Teor.Fiz. 23, 236 (1952)
/3/ D.R.Tilley, Proc.Phys.Soc. 85, 1177 (1965)
/4/ L.P.Gor'kov and T.K.Melik-Barkhudarov, Soviet Phys. JETP 18, 1031 (1964)
/5/ C.Caroli, P.G. de Gennes, J.Matricon, Physik d. kondensierten Materie 1, 176 (1963)
/6/ P.C.Hohenberg, N.R.Werthamer, Phys.Rev. 153, 493 (1967)
/7/ L.P.Gor'kov, Soviet Phys. JETP 10, 998 (1960); Soviet Phys. JETP 10, 593 (1960)
/8/ W.A.Reed, E.Fawcett, P.P.M.Meincke, P.C.Hohenberg, N.R.Werthamer, Proc. LT10, Moscow 1966, ed. by M.P.Malkov, vol. IIA, p. 288
/9/ K.Takanaka, T.Nagashima, Progr.Theor.Phys. 43, 18 (1970)
/10/ T.Nagashima, Progr.Theor.Phys. 47, 37 (1972)
/11/ H.Teichler, phys.stat.sol.(b) 69, 501 (1975)
/12/ F.C. von der Lage, H.A.Bethe, Phys.Rev. 71, 612 (1947)
/13/ C.E.Gough, Solid State Comm. 6, 215 (1968)
/14/ I.Williams, A.M.Court, Solid State Comm. 7,169 (1969)
/15/ D.E.Farrell, B.S.Chandrasekhar, S.Huang, Phys.Rev. 176, 562 (1968)
/16/ S.J.Williamson, Phys.Rev. B2, 3545 (1970)
/17/ M.Yamamoto, N.Ohta, T.Ohtsuka, J.Low Temp.Phys. 15, 231 (1974)

/18/ R.J.Hembach, F.K.Mullen, R.W.Genberg, Proc. LT14, Helsinki 1975, ed. by M.Krusius and M.Vuorio (North Holland, Amsterdam and American Elsevier, New York) vol.2, p.313
/19/ T.Nagashima, H.Fukuyama, to be published and private communication
/20/ N.Ohta, Doctoral Thesis, Tohoku University, 1975
/21/ D.E.Farrell, B.S.Chandrasekhar, H.V.Culbert, Phys.Rev. $\underline{177}$, 694 (1969)
/22/ N.Ohta, N.Tanaka, T.Ohtsuka, Phys.Lett. $\underline{49A}$, 363 (1974)
/23/ T.Nagashima, private communication
/24/ M.Roger, R.Kahn, J.M.Delrieu, Phys.Lett. $\underline{50A}$, 291 (1974); cf. also M.Roger, Thesis, University of Paris VI, June 1974
/25/ H.Kiessig, U.Essmann, H.Teichler, W.Wiethaup, Phys.Lett. $\underline{51A}$, 333 (1975)
/26/ S.J.Williamson, L.Valby, Phys.Rev.Lett. $\underline{24}$, 1061 (1970); cf. also S.J.Williamson, ref.16
/27/ K.Takanaka, Progr.Theor.Phys. $\underline{46}$, 357 (1971)
/28/ K.Maki, T.Tsuzuki, Phys.Rev. $\underline{139}$, A 868 (1965)
/29/ K.Takanaka, private communication
/30/ E.Helfand, N.R.Werthamer, Phys.Rev. $\underline{147}$, 288 (1966)
/31/ T.Ohtsuka, Y.Kimura, Physica $\underline{55}$, 562 (1971)
/32/ S.J.Williamson, private communication
/33/ L.F.Mattheiss, Phys.Rev. $\underline{B1}$, 373 (1970)
/34/ K.Takanaka has recently pointed out that HW's expression for $\overline{h^*(0)}$ must be corrected, cf.paper C-5, this volume, p.75
/35/ K.Takanaka, A.Hubert, Proc. LT14, Helsinki 1975, ed. by M.Krusius and M.Vuorio (North Holland, Amsterdam and American Elsevier, New York) vol.2, p.309
/36/ B.Obst, phys.stat.sol. (b) $\underline{45}$, 453 (1971); ibid. 467

C-1   THE INFLUENCE OF FERMI SURFACE ANISOTROPY ON $H_{c2}(T)$

P. Entel*

Institut für Theoretische Physik der Universität zu Köln

D-5 Köln 41, W-Germany

M. Peter

Département de Physique de la Matière Condensée

Université de Genève, CH-1211 Genève 4, Switzerland

Abstract

We investigate the influence of Fermi surface anisotropy on the upper critical field $H_{c2}(T)$ of bulk type-II superconductors within the framework of an effective two-band model for the electron gas. It is found that the temperature dependence of H_{c2} shows typical departures from the corresponding curve for a single band isotropic superconductor. Two types of parameters determine these departures:
(1) The signs and the strengths of the interband electron-phonon coupling constants determine the enhancement of $H_{c2}(0)$ with respect to the Helfand-Werthamer value for the isotropic case.
(2) The interband transport lifetimes τ_{ij}^{tr} and interband Coulomb scattering times τ_{ij} (i,j label the states in different bands or in different zones of the Fermi surface) reduce the enhanced

---

*Work performed within the research program of the Sonderforschungsbereich 125-Aachen/Jülich/Köln.

$H_{c2}(0)$-values.

By choosing reasonable values for the coupling constants and for the lifetimes we find that an enhancement of the $H_{c2}(0)$-values by more than 100% is possible /1/. We can thus explain the difference of 30% between measured values of $H_{c2}(0)$ and the prediction of weak coupling theories, which has been observed in the system $Nb_{75\pm x}Pt_{25\mp x}$ and $Cs_{0.1}WO_{2.9}F_{0.1}$ recently /2/.

1. Introduction

We distinguish between two different anisotropy effects:
(1) Crystalline anisotropy leads to an upper critical field which depends on the orientation of the crystal relative to the direction of the applied field, i.e. $H_{c2}(T) = H_{c2}(T,\varphi,\theta)$. In polycrystalline or in very dirty samples this effect should vanish.
(2) Fermi surface anisotropy or the existence of two or more conduction bands leads in pure samples to enhanced $H_{c2}(T)$ values with respect to the corresponding $H_{c2}(T)$ values of a single band isotropic superconductor /3/. In dirty samples this anisotropy effect is diminished.

In the following we concentrate on the second effect and show that the experimental $H_{c2}(T,\varphi,\theta)$ values (for fixed angles φ and θ) in $Cs_{0.1}WO_{2.9}F_{0.1}$ /2/ are compatible with $H_{c2}(T)$ calculations within the framework of a simple two-band model. Moreover, experimental data for different angles φ and θ can be used to construct the Fermi surface.

For a generalization to the case of N bands and an inclusion of strong coupling effects we refer to ref./1/.

2. Two-Band Model

Considering a pairing interaction between particles having the same band index we have to solve the coupled equations for the order parameter

$$\Delta_i^+(\underline{x}) = g_{ii} \frac{1}{\beta\hbar^2} \sum_n \int d^3\underline{x}' <G_i^o(\underline{x}'\underline{x},-n)G_i^o(\underline{x}'\underline{x},n) \Delta_i^+(\underline{x}')>$$

$$+ g_{ij} \frac{1}{\beta\hbar^2} \sum_n \int d^3\underline{x}' <G_j^o(\underline{x}'\underline{x},-n)G_j^o(\underline{x}'\underline{x},n) \Delta_j^+(\underline{x}')> \quad (1)$$

$$i,j = 1,2$$

Here $G_i^o(\underline{xx}',n)$ is the one particle Green's function for the normal state containing the magnetic field

$$G_i^o(\underline{xx}',n) = -\frac{m_i}{2\pi\hbar^2} \frac{1}{|\underline{x}-\underline{x}'|} \exp\{i(k_{Fi} \text{sgn}(n) - \frac{|\omega_n|}{v_{Fi}}|\underline{x}-\underline{x}'|\}$$

$$\cdot \exp\{i \frac{e}{c\hbar} \int_{\underline{x}'}^{\underline{x}} \underline{A}(\underline{s}) d\underline{s}\} \quad (2)$$

g_{ij} are the electron-phonon coupling constants and the brackets < > denote averaging over impurity configurations.

The general solution of (1) is very complicated. In the dirty limit and retaining s- and p-wave scattering, Eq.(1) transforms into

$$\Delta_i^+(\underline{x}) = g_{ii} \frac{1}{\beta\hbar^2} \sum_n \int d^3\underline{x}' \tilde{G}_i^o(\underline{x}'\underline{x},-\tilde{n}_i) \tilde{G}_i^o(\underline{x}'\underline{x},\tilde{n}_i) \tilde{\Delta}_i^+(\underline{x}')$$

$$+ g_{ij} \frac{1}{\beta\hbar^2} \sum_n \int d^3\underline{x}' \tilde{G}_j^o(\underline{x}'\underline{x},-\tilde{n}_j) \tilde{G}_j^o(\underline{x}'\underline{x},\tilde{n}_j) \tilde{\Delta}_j^+(\underline{x}') \quad (3)$$

where the renormalized frequencies and order parameters are given by:

$$\hbar\tilde{\omega}_{ni} = \hbar\omega_n + \text{sgn}(n)\left\{\frac{\hbar}{2\tau_{ii}} + \frac{\hbar}{2\tau_{ij}}\right\} \tag{4}$$

$$\frac{\hbar}{\tau_{ij}} = c\pi N_j(0) \int d\Omega \, |V_{ij}(\theta)|^2 \tag{5}$$

and

$$\tilde{\Delta}_i^+(q) = \{(1-\hat{I}_{11})(1-\hat{I}_{22}) - \hat{I}_{12}\hat{I}_{21}\}^{-1}\{(1-\hat{I}_{jj})\Delta_i^+(q) + \hat{I}_{ij}\Delta_j^+(q)\} \tag{6}$$

The operators $\hat{I}_{ij}$ are known /4,5/:

$$\hat{I}_{ij}(q,\tilde{n}_j) = \frac{1 - \frac{1}{3}\tau_{ij}\tau_{ij}^{tr} v_{Fj}^2 (q - \frac{2e\mathbf{A}}{\hbar c})^2}{2\tau_{ij}|\hbar\tilde{\omega}_{nj}|} \tag{7}$$

with

$$\frac{\hbar}{\tau_{ij}^{tr}} = c\pi N_j(0) \int d\Omega \, |V_{ij}(\theta)|^2 (1 - \cos\theta)$$

Solving (3) in analogy to the one-band case /4/ we obtain finally:

$$\ln\frac{T_c}{T} = \frac{1}{2}\{K_+ + K_- \mp A \pm \sqrt{(K_+ - K_-)^2 + A^2 - 2B(K_+ - K_-)}\} \tag{8}$$

where the lower sign refers to the case $\lambda_{11}\lambda_{22} - \lambda_{12}\lambda_{21} > 0$, $\lambda_{ij} = g_{ij}N_j(0)$.

$$K_\pm = \Psi(\frac{1}{2} + \tilde{P}_\pm) - \Psi(\frac{1}{2})$$

$$\tilde{P}_\pm = \frac{1}{2\pi k_B T} \frac{1}{2}\{\alpha_1 + \alpha_2 \pm \sqrt{(\alpha_1 - \alpha_2)^2 + 4\beta_1\beta_2}\}$$

$$\alpha_i = \frac{\hbar}{2\tau_{ij}} + \frac{\hbar}{3}\tau_{ii}^{tr} v_{Fi}^2 \frac{eH_{c2}}{\hbar c}$$

$$\beta_i = \frac{\hbar}{2\tau_{ij}} - \frac{\hbar}{3}\tau_{ij}^{tr} v_{Fj}^2 \frac{eH_{c2}}{\hbar c}$$

$$A = \frac{\sqrt{(\lambda_{11}-\lambda_{22})^2 + 4\lambda_{12}\lambda_{21}}}{|\lambda_{11}\lambda_{22} - \lambda_{12}\lambda_{21}|}$$

$$B = \frac{(\alpha_1-\alpha_2)(\lambda_{11}-\lambda_{22}) - 2\lambda_{12}\beta_2 - 2\lambda_{21}\beta_1}{(\lambda_{11}\lambda_{22}-\lambda_{12}\lambda_{21})\sqrt{(\alpha_1-\alpha_2)^2 + 4\beta_1\beta_2}}$$

Comparing (8) with the corresponding result for a two-band superconductor containing magnetic impurities /6/ we see the equivalence of different pair breaking mechanisms.

3. The System $Cs_{0.1}WO_{2.9}F_{0.1}$

$Cs_{0.1}WO_{2.9}F_{0.1}$ belongs to the tungsten fluoroxide bronze. The Cs and F atoms, distributed along hexagonal channels surrounded by WO_6 octahedra, are expected to contribute their valence electrons to the empty WO_3 conduction band. No band structure calculation is available. It is believed, however, that the conduction band is of strong d-band character and is split by the crystalline field into two d sub-bands (Fig.1) /7/. This enables us to treat this system as a two-band superconductor.

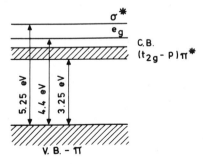

Fig.1: Electron energy band structure of WO_3 after /7/

4. The Upper Critical Field of $Cs_{0.1}WO_{2.9}F_{0.1}$

$H_{c2}(T,\varphi,\theta)$ measurements by Skokan et al. /2/ reveal that both anisotropy effects discussed in the introduction are present. Moreover, they depend strongly on the degree of disorder in this system. The simplest approach is to describe the degree of disorder by effective transport scattering times.

For a given sample orientation (case a in Fig.2 of ref. /2/) Fig.2 shows the fit of theoretical $H_{c2}(T)$-values evaluated according to Eq.(8) to the experimental results. The dashed curves (denoted by $H_{c2}^1(T)$ and $H_{c2}^2(T)$) are obtained in the following way. We assume two uncorrelated bands (no electron-phonon coupling between the two bands and no interband scattering by impurities): The slopes

$$\left. \frac{dH_{c2}^i}{dT} \right|_{T=T_c}$$

of the $H_{c2}^i(T)$ curves are taken from the experimental data (sample 1

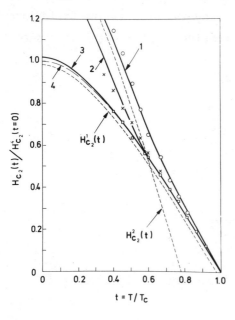

Fig.2: Comparison of experimental $H_{c2}(T)$-values /2/ for a fixed sample orientation with fitted theoretical curves 1 to 3. The experimental results (open circles, crosses and squares) correspond to three different samples with different degrees of crystalline order.

of ref. /2/). For a nonvanishing weak electron-phonon coupling we obtain one single $H_{c2}(T)$ curve (curve 1) for both bands which fits well the experimental points (open circles).

With increasing disorder we obtain two competing effects:
(1) Smaller <u>intraband</u> and <u>interband transport scattering</u> times <u>enhance</u> $H_{c2}(T)$.
(2) Smaller <u>interband Coulomb scattering</u> times will <u>decrease</u> $H_{c2}(T)$. In general, the latter effect will dominate (this shows the importance of having s <u>and</u> p wave scattering).

For nonzero interband Coulomb scattering we obtain the curves 2 and 3: It can be seen, that the reduction of T_c is negligibly small whereas $H_{c2}(0)$ is reduced by $\approx 50\%$ ($H_{c2}(0)|_{\text{curve 1}} \approx 2$, $H_{c2}(0)|_{\text{curve 3}} \approx 1$). The following parameters have been used: The temperature is normalized by T_c and $H_{c2}(T)$ is normalized by $H_{c2}^4(t=0)$, which is the limiting value for the isotropic case.

Taking the Debye temperature to be $\theta_D = 589$ K /8/, we obtain (with the prescription of taking the right slope $\left.\frac{dH_{c2}}{dT}\right|_{T_c}$):
$\lambda_{11} = 0.2$, $\lambda_{22} = 0.194$, $\lambda_{12} = -0.0024$, $\lambda_{21} = -0.0051$.
Furthermore, the slope determines the diffusion constants D_{ij}, which are defined by

$$D_{ij} = \frac{1}{3} \tau_{ij}^{tr} v_{Fj}^2 \frac{e}{c}$$

We get $D_{11} = 0.473 \times 10^{-4}$ eV/kG and $D_{22} = 0.226 \times 10^{-4}$ eV/kG where we have used the single-band results

$$\left.\frac{dH_{c2}^i(T)}{dT}\right|_{T=T_c} = -\frac{4k_B}{\pi}\frac{1}{D_{ii}}$$

$$H^i_{c2}(0) = \frac{k_B T^i_c}{1.13} \frac{1}{D_{ii}}$$

The interband scattering times cannot be determined uniquely. We believe them to be of the same order of magnitude as the intraband scattering times and choose:

$$D_{ij} = 0 \quad (i \neq j)$$

$$\frac{\hbar}{2\tau_{21}} = 3.2 \times 10^{-5} \text{ eV}, \quad \frac{\hbar}{2\tau_{12}} = 1.53 \times 10^{-5} \text{ eV} \quad \text{for curve 2}$$

$$\frac{\hbar}{2\tau_{21}} = 1.25 \times 10^{-3} \text{ eV}, \quad \frac{\hbar}{2\tau_{12}} = 0.59 \times 10^{-3} \text{ eV} \quad \text{for curve 3}$$

One final remark: Instead of using (8) we could use the pure limit result when fitting the two different slopes to the experimental data:

$$\left. \frac{dH^i_{c2}}{dT} \right|_{T=T_c} = - \frac{6 c k_B^2 4\pi^2}{7\zeta(3)\hbar e T^i_c} \frac{1}{v_{Fi}} \left(\frac{T_{ci}}{v_{Fi}}\right)^2$$

which gives

$$v_{Fi} \simeq \alpha_i \times 10^7 \text{ cm s}^{-1}, \quad \alpha_i < 1$$

and indicates, that $Cs_{0.1}WO_{2.9}F_{0.1}$ has very flat energy bands and a high density of states. Using now all different v_i-data thus obtained for different φ and θ one could get an impression of what the Fermi surface looks like.

References

/1/ P.Entel, M.Peter, J.Low Temp.Phys. 22, 613 (1976)
/2/ M.R.Skokan, R.C.Morris, W.G.Moulton, Phys.Rev. B13, 1077 (1976)
/3/ E.Helfand, N.R.Werthamer, Phys.Rev.Lett. 13, 686 (1964)
/4/ K.Maki, Physics 1, 21 (1964)
/5/ W.S.Chow, Phys.Rev. 176, 525 (1968)
/6/ P.Entel, W.Klose, O.Fischer, G.Bongi, Z.Physik B21, 363 (1975)
/7/ S.K.Deb, Phil.Mag. 27, 801 (1973)
/8/ R.W.Vest, M.Griffel, J.F.Smith, J.Chem.Phys. 28, 293 (1958).

C-2 IMPURITY DEPENDENCE OF H_{c2} ANISOTROPY IN NIOBIUM

E.Seidl and H.W.Weber

Atominstitut der Österreichischen Universitäten

A-1020 Vienna, Austria

1. Introduction

The anisotropy of the upper critical field H_{c2} in cubic type-II superconductors is well established, e.g. /1/, and rather detailed experimental information on this effect is available, at least for high purity materials. For impure samples, however, only a few investigations have been reported so far /2,3/.

Based on the neutron diffraction results by Weber et al. /4/, who observed that in single crystalline niobium containing a considerable amount of interstitially dissolved nitrogen and even some 10^9 normalconducting precipitates per cm^3 a distinct correlation between the flux line lattice and the crystal lattice was still maintained (cf. also /5/), we decided to study the dependence of the anisotropy effect on the impurity concentration systematically. To do this, the impurity parameter of one and the same niobium crystal was increased in small steps and the anisotropy of H_{c2} measured at each stage in the temperature range from 1.8 K to the transition temperature.

In the light of recent theories /6/ a combined investigation of the temperature and impurity dependence of H_{c2} anisotropy as will be reported in the following sections should allow to separate the contributions of various physical mechanisms leading to the observed anisotropy effect and should, therefore, help to clarify whether the anisotropy is simply a normal state effect (through the anisotropic properties of the Fermi surface, the magnitude and direction of the Fermi velocity and the density of states) or involves an anisotropic electron-phonon interaction. This analysis of our data is presented in the following paper /7/.

2. Experimental

In order to study the impurity dependence of the anisotropy effect, we chose the system niobium - nitrogen, where nitrogen was allowed to diffuse into a high-purity niobium single crystal (Nb 0) of 8 mm diameter and 40 mm length and having a <110> crystal direction parallel to its cylinder axis.

Starting from ultra high vacuum conditions the diffusion process was made at a temperature of 1900 °C and successively increasing nitrogen pressures ranging from 1×10^{-6} torr to 4×10^{-4} torr /8/. In order to ensure a homogeneous distribution of the interstitially dissolved nitrogen atoms diffusion times of more than 30 hours were used.

After each diffusion step the axial (<110>) magnetization, the residual resistivity, the transition temperature and the H_{c2} anisotropy were measured. The magnetization curves proved, that almost ideal reversibility of the magnetization was achieved indicating the desired interstitial dissolution of the nitrogen atoms and the corresponding reduction of the mean free path except for the last step (Nb6) involving the highest nitrogen concentration, where precipitation of normalconducting Nb_2N precipitates /9/ occured (cf.

Impurity Dependence of H_{c2} Anisotropy

section 3). An evaluation of the magnetization curves with respect to the Ginzburg-Landau parameter κ at $T = T_c$ and independent measurements of the residual resistivities enable a comparison with the prediction of the Gor'kov-Goodman relation. The result shown in Fig.1 confirms the expected linear increase of κ and H_{c2} with increasing resistivity (or decreasing mean free path). This is, of course, no longer true for sample Nb6 (not shown in the figure), where the precipitate formation reduces the amount of nitrogen dissolved on interstitial positions leading to a reduction of the mean free path, which falls considerably below the value corresponding to the real nitrogen content.

Inductive measurements of the transition temperature showed a gradual decrease of T_c with increasing impurity content. This variation is, however, rather small ($\sim 4\%$ for the range of impurity parameters 0.025 to 1.77). The system niobium - nitrogen seems, therefore, to be more suitable for the present purpose than the Nb-Ta alloy system studied previously /3/ (cf. also ref./1/).

Some characteristic parameters of our system may be summarized as follows:

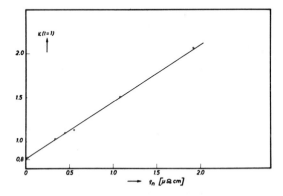

Fig.1: Dependence of the Ginzburg-Landau parameter on the normal state resistivity deduced from magnetization and residual resistivity measurements (ooo) compared with the Gor'kov-Goodman relation (———)

residual resistivity ratios: 595, 48.1, 37.0, 29.6, 15.7, 9.4, 4.6
impurity parameters ($\alpha = 0.882\ \xi_0/\ell$): 0.025, 0.31, 0.41, 0.52, 0.99, 1.77, 4.1
nitrogen contents (10^{-2} at%): 0, 11.3, 13.0, 14.5, 27.5, 55.8, 116.3
$\Delta\kappa(t=1)$: 2.22/at%, $\Delta\mu_0 H_{c2}^{<110>}$ (4K) = 650 mT/at%, $\Delta\rho(293)$ = 4.0 $\mu\Omega$ cm/at%, $\Delta\rho(4.2)$ = 3.5 $\mu\Omega$ cm/at%.

Finally, the measurements of H_{c2} anisotropy were made /10/ by applying a static magnetic field provided by a superconducting Helmholtz pair perpendicular to the cylinder axis and rotating the crystal within a (110) plane. The transition at H_{c2} was obtained by imposing a small ripple field parallel to the cylinder axis and recording the pick-up signal with a lock-in amplifier. At each stage of the diffusion process the angular dependence of H_{c2} was measured at about 25 positions between the <100> and the <110> directions, about 20 - 30 fixed temperatures were used to determine its temperature dependence. The accuracy of H_{c2} concerning its angular dependence lies within ± 0.1 mT, whereas the absolute values of H_{c2} are considerably more uncertain due to the extrapolation procedure; the temperature stability was always within ± 1 mK.

3. Results

Typical experimental results for the purest sample (Nb 0) are shown in Fig.2, the extrapolation procedure used to determine H_{c2} is indicated in the insert. In order to analyze the data, orthonormalized cubic harmonic functions as defined by Von der Lage et al. /11/ were used:

$$H_{c2}(T,\alpha,\beta_i) = \sum_{l=0,4,6,8} A_l(T,\alpha)\, H_l(\beta_i) \tag{1}$$

where the β_i denote the direction cosines of the external field with respect to the cubic axes. Least mean square fits of the data showed

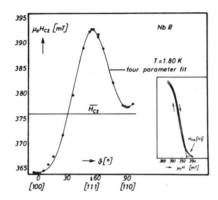

Fig.2: Typical experimental result of H_{c2} anisotropy ($\overline{H_{c2}} = A_0$)

that cubic harmonic functions up to the order $l = 8$ were needed to describe the angular dependence of H_{c2} adequately.

In order to present a summary of the experimental results emphasizing the influence of impurity concentration on the anisotropy effect, Figs.3 and 4 show a plot of the reduced coefficients A_l of Eq.(1) ($a_4 = A_4/A_0$ and $a_6 = A_6/A_0$) versus the impurity parameter α with the temperature T as parameter. (Although we are aware of the slight inconvenience resulting from the choice of the actual measuring temperature T as parameter instead of the reduced temperature T/T_c, the impurity effect is demonstrated more clearly by the presentation used. Furthermore, a plot of the coefficients a_4 and a_6 versus reduced temperature t and with α as parameter is shown in the following paper /7/.)

It will be noted from Fig.3, that the coefficient a_4 decreases sharply with the addition of only a small amount of impurities and more slowly furtheron. For comparison the results on sample Nb6 showing a rather homogeneous distribution of normalconducting Nb_2N precipitates ($\sim 4 \times 10^9$ cm^{-3}) and considerable bulk pinning are in-

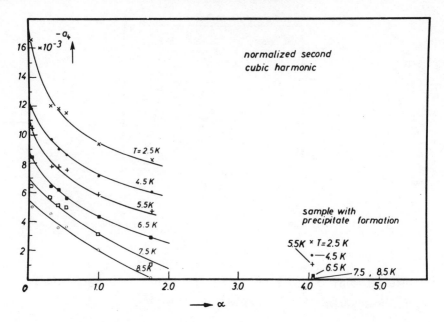

Fig.3: Coefficient $a_4 = A_4/A_0$ versus impurity parameter α

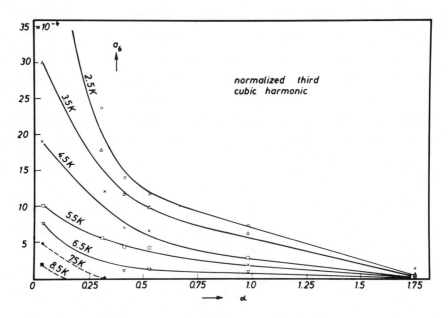

Fig.4: Coefficient $a_6 = A_6/A_0$ versus impurity parameter α

cluded in Fig.3. Despite of the presence of pinning forces, which tend to destroy correlations with the crystal lattice, a distinct anisotropy effect is still observed. A rough estimate of the critical current density from the magnetization curve shows, that it lies below the limit for a total destruction of correlations deduced by Schelten from neutron diffraction experiments (Fig.5 of ref./5/). A detailed investigation of the pinning forces in this sample is under way /12/.

Finally, the impurity dependence of the coefficient a_6 is shown in Fig.4. The decrease of a_6 with the addition of small impurity concentrations is much more pronounced than with a_4 (e.g. at $T=2.5$ K a_6 drops from 53×10^{-4} to 24×10^{-4} upon adding 11.3×10^{-2} at% nitrogen). The same holds for higher impurity concentrations, a_6 becoming negligibly small at an impurity parameter $\alpha = 1.77$.

The analysis of these results in terms of a recent microscopic theory /6/ and especially the consequences of the different response of the coefficients a_4 and a_6 to the addition of impurities are presented in the following paper.

References

/1/ T.Ohtsuka, paper R-2, this volume, p. 27
/2/ W.A.Reed, E.Fawcett, P.P.M.Meincke, P.C.Hohenberg, N.R.Werthamer, Proc. LT 10 (M.P.Malkov, Ed.), Viniti 1967, vol. IIa, p. 288
/3/ M.Yamamoto, N.Ohta, T.Ohtsuka, J.Low Temp.Phys. <u>15</u>, 231 (1974)
/4/ H.W.Weber, J.Schelten, G.Lippmann, J.Low Temp.Phys. <u>16</u>, 367 (1974)
/5/ J.Schelten, paper R-4, this volume, p. 113
/6/ H.Teichler, paper R-1, this volume, p. 7
/7/ H.Teichler, paper C-3, this volume, p. 65
/8/ E.Seidl, H.W.Weber, to be published

/9/ G.Antesberger, H.Ullmaier, Phil.Mag. 29, 1101 (1974)
/10/ E.Seidl, H.W.Weber, H.Teichler, to be published
/11/ F.C.Von der Lage, H.A.Bethe, Phys.Rev. 71, 612 (1947)
/12/ N.Giesinger, E.Seidl, H.W.Weber, to be published

C-3 MICROSCOPIC ANISOTROPY PARAMETERS OF Nb

H. Teichler, Institut für Physik am Max-Planck-Institut für Metallforschung and Institut für Theoretische und Angewandte Physik der Universität, D-7 Stuttgart, W-Germany

The anisotropy of the upper critical field H_{c2} (e.g. /1/) reveals the microscopic anisotropies of the superconducting electron system, i.e., the anisotropies of the Fermi surface, of the Fermi velocity, and of the attractive electron-electron interaction. According to a recent nonlocal theory /2,3/ of H_{c2} anisotropy in cubic superconductors with low or moderate impurity content at arbitrary temperatures below T_c, the various microscopic anisotropies contribute to the anisotropy of H_{c2} with different temperature and impurity dependent weights. Due to this fact investigations of H_{c2} anisotropy may be used to estimate the magnitudes of these various anisotropies in the superconducting electron system. - Details of the theory are reviewed in /3/. Here we use this theory to analyze experimental data of Seidl and Weber /4/ on the H_{c2} anisotropy in the system Nb-N.

By inverting the relation

$$\frac{H_{c2}(\underline{e}_H) - \overline{H}_{c2}}{\overline{H}_{c2}} = \sum_{l=4,6,8} H_l(\underline{e}_H) a_l(T/T_c, \alpha)$$

($\overline{H}_{c2}$: average of H_{c2} over all field directions $\underline{e}_H$; $H_l(\underline{e}_H)$: cubic harmonics /5/) Seidl and Weber /4/ determined the expansion coefficients $a_1(T/T_c,\alpha)$ on six Nb-N samples with different values of the impurity parameter $\alpha = \hbar/(2\pi k_B T_c \tau)$. From these coefficients the microscopic anisotropy parameters $\gamma_l^{(i)}$ defined in /2,3/ were deduced by adapting, e.g., Eq.(12) of /3/ to these data by least mean squares fits. The experimental values /4/ of $a_4(T/T_c,\alpha)$ and $a_6(T/T_c,\alpha)$ and the analytical curves with parameters $\gamma_l^{(i)}$ presented in Table I are shown in Figs.1 and 2. (For comparison some data points obtained by Williamson /1/ are also included in these figures.)

Table I: Microscopic anisotropy parameters of Nb

	$\gamma_l^{(1F)}$	$\gamma_l^{(1\varphi)}$	$\gamma_l^{(2)}$
l = 4	-0.177	-0.062	-0.103
l = 6	-0.042	0.251	-0.036

According to Table I the main contributions to $a_4(T/T_c,\alpha)$ result from $\gamma_4^{(1F)}$ and $\gamma_4^{(2)}$, i.e. the l = 4 component of H_{c2} anisotropy is mainly due to electron band-structure anisotropies of the normal state. In $a_6(T/T_c,\alpha)$ the parameter $\gamma_6^{(1\varphi)}$ gives the overwhelming contribution, i.e., in <u>pure</u> Nb the main part of the l = 6 component of H_{c2} anisotropy originates from anisotropies in the attractive electron-electron interaction, which reveal effects of the electron-phonon interaction, phonon dispersion-curves, and Coulomb pseudopotential.

Up to now the microscopic anisotropy parameters have not yet been determined by other methods or calculated from first principles. However, according to a recent theory /3,6/ the phenomenon of preferential orientations of the flux line lattice (FLL) relative to the crystal lattice reflects the $\gamma_l^{(i)}$, too. Thus, in order to test the reliability of the $\gamma_l^{(i)}$ deduced from H_{c2} anisotropy we may use these values to estimate the preferential orientation of the FLL in Nb.

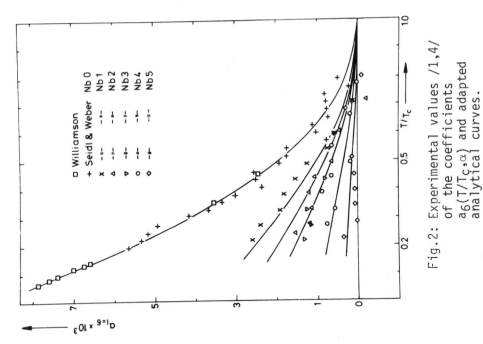

Fig.2: Experimental values /1,4/ of the coefficients $a_6(T/T_c, \alpha)$ and adapted analytical curves.

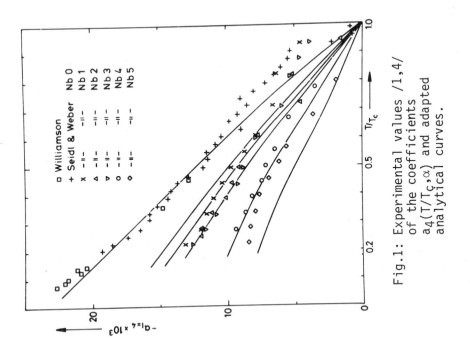

Fig.1: Experimental values /1,4/ of the coefficients $a_4(T/T_c, \alpha)$ and adapted analytical curves.

If we assume a triangular FLL and use the $\gamma_1^{(i)}$ values of Table I, the theory /6/ predicts for magnetic fields parallel to the <111>, <110>, or <112> crystal axes, that at low temperatures one of the basis vectors of the FLL should be parallel to the <1$\bar{1}$0> direction. In addition, if for magnetic fields parallel to the <001> crystal axis a square FLL is realized, the theory /6/ predicts that the basis vectors of the FLL are parallel to the crystallographic <110> directions. These theoretical results are in agreement with experimental observations /7 - 9/.

The author is indebted to Professor Dr.A.Seeger for his stimulating interest in this problem. He wishes to thank Drs.E.Seidl and H.W. Weber for providing their experimental data prior to publication.

References

/1/ S.J.Williamson, Phys.Rev. B2, 3545 (1970)
/2/ H.W.Pohl, H.Teichler, phys.stat.sol.(b) 75, 205 (1976)
/3/ H.Teichler, paper R-1, this volume, p.7
/4/ E.Seidl, H.W.Weber, paper C-2, this volume, p.57
/5/ F.C. von der Lage, H.A.Bethe, Phys.Rev. 71, 612 (1947)
/6/ K.Fischer, H.Teichler, submitted to Phys.Lett.
/7/ J.Schelten, G.Lippmann, H.Ullmaier, J.Low Temp.Phys. 14, 213 (1974)
/8/ R.Kahn, G.Parette, Solid State Commun. 13, 1839 (1973)
/9/ U.Essmann, Physica 55, 83 (1971).

C-4 ANISOTROPY OF H_{c2} IN PbTl-ALLOYS

H.Kiessig, U.Essmann, H.Teichler, W.Wiethaup

Institut für Physik am Max-Planck-Institut für Metallforschung and Institut für Theoretische und Angewandte Physik der Universität, D-7 Stuttgart, W-Germany

We studied the anisotropy of the upper critical field H_{c2} for a Pb-1.8 at% Tl alloy in the temperature range between 1.8 K and 4.2 K. This composition is of special interest, since it shows a transition from type-I to type-II superconductivity at 4.6 K /1/.

Monocrystalline samples were grown by the Bridgman technique /2/. The magnetization was measured with a commercially available vibrating-sample magnetometer /3/. The samples were mounted in a ^{4}He tail cryostat between the pole faces of a conventional electromagnet in a way, which allowed the variation of field direction in the (110) plane of the sample by simply rotating the magnet. The field strength was measured by a Hall-effect gaussmeter. During the experiment temperature fluctuations due to variations of the helium vapor pressure were less than ±1 mK.

The magnetization near H_{c2} was plotted on an x-y recorder. At constant temperature the slopes of the magnetization curves near H_{c2} were found to be independent of crystal orientation. This indicates that $\kappa_2 = f(dM/dH|_{H_{c2}})$ is isotropic. For a given orientation of the magnetic field we define H_{c2} by the intercept of the tangents at the magnetization curve above and below the kink in the magnetization near H_{c2}.

The orientation dependence of H_{c2} may be described in terms of orthonormalized cubic harmonics according to

$$H_{c2}(\underline{e},T) = \sum_n A_n(t) H_n(\underline{e}) \tag{1}$$

Here $\underline{e} = (e_x, e_y, e_z)$ denotes the unit vector in the direction of the applied field and $t = T/T_c$ is the reduced temperature. The cubic harmonics introduced by Von der Lage and Bethe /4/ are

$$H_1 = 1$$
$$H_2 = 5.73 \, (e_x^4 + e_y^4 + e_z^4 - 0.6)$$
$$H_3 = 147.2 \, (e_x^2 e_y^2 e_z^2 + 0.00794 \, H_2 - 0.0095)$$

The coefficients $A_n(t)$ were determined by fitting (1) (up to the term with n = 3) to the experimental data points by a least-mean-squares procedure. The quality of the fit may be seen from Fig.1. The temperature dependence of the coefficients $A_n(t)$ is shown in

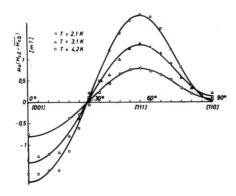

Fig.1: Orientation dependence of $\mu_o(H_{c2} - \bar{H}_{c2})$ for three different temperatures. Full curves obtained by least-mean-squares fits

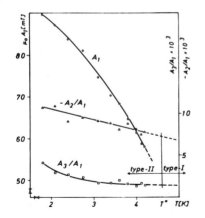

Fig.2:
Temperature dependence of the coefficients of the cubic harmonics

Fig.2. $A_1(t) = \overline{H}_{c2}(t)$ describes the average critical field. The reduced coefficient $A_2(t)/A_1(t)$ shows a linear temperature dependence, whereas according to Fig.2 the temperature dependence of $A_3(t)/A_1(t)$ is non-linear.

A comparison of our results with those of Williamson /5/ on pure niobium gives

$$\frac{A_2(t)/A_1(t)\big|_{\text{Pb-1.8 at\% Tl}}}{A_2(t)/A_1(t)\big|_{\text{Nb}}} \simeq 0.6$$

and

$$\frac{A_3(t)/A_1(t)\big|_{\text{Pb-1.8 at\% Tl}}}{A_3(t)/A_1(t)\big|_{\text{Nb}}} \simeq 1$$

The anisotropy of H_{c2} in Pb-1.8 at% Tl is thus of the same order of magnitude as in Nb.

Our experiments indicate that $A_2(t)/A_1(t)$ and $A_3(t)/A_1(t)$ remain finite at the transition from type-II to type-I superconductivity. The transition temperature T* may be defined by the equation

$$H_{c2}(T^*,\underline{e}) \equiv H_c(T^*) \tag{2}$$

where H_c denotes the thermodynamic critical field, which is independent of orientation. $T^* = T^*(\underline{e})$ is anisotropic if $A_n \neq 0$ ($n = 2, 3, \ldots$). For small anisotropies $T^*(\underline{e})$ is related to the anisotropy of $H_{c2}(\underline{e})$

$$\frac{T^*(\underline{e}) - \overline{T}^*}{\overline{T}^*} = \frac{\Delta H_{c2}(\underline{e}, \overline{T}^*)}{\overline{H}_{c2}(\overline{T}^*)} \left[\frac{T}{\overline{H}_{c2}} \left(\frac{\partial H_{c2}}{\partial T} - \frac{\partial H_c}{\partial T} \right) \right]^{-1} \Big|_{T = \overline{T}^*} \quad (3)$$

where $\overline{T}^*$ denotes the average of $T^*(\underline{e})$ over all space directions.

Equation (3) shows that around $\overline{T}^*$ an interval ΔT^* exists where the orientation of the applied field determines whether the superconductor will show type-II or type-I behavior. Figure 3 illustrates the orientation dependence of this transition, if supercooling and demagnetizing effects are disregarded. At a given temperature T, $H_{c2}(\underline{e})$ can be measured only for a certain range of orientations (between $\underline{e}_1$ and $\underline{e}_2$). For the remaining orientations the superconductor shows type-I behavior and the experiments give the thermodynamic critical field H_c. Figure 4 illustrates that the anisotropy of H_{c2} and the thermodynamic condition, that the area under the

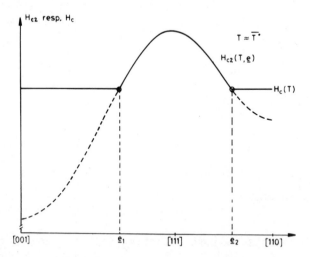

Fig.3: Orientation dependence of critical fields for a superconductor at a temperature T near T^* in the absence of supercooling and demagnetizing effects. The dashed curve shows $H_{c2}(\underline{e})$ as it may be obtained from supercooling experiments.

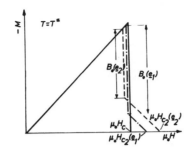

Fig.4: Ideal magnetization curves of an anisotropic superconductor at a temperature near T* for two different orientations $\underline{e}_1$ and $\underline{e}_2$ of the external magnetic field

magnetization curve at a given temperature has to be independent of crystal orientation, together with the observed isotropy of κ_2 lead to an anisotropy of the lower critical field H_{c1} and of the spontaneous induction B_o at H_{c1} /1/.

The authors are indebted to Professor A.Seeger for critical comments on this work.

References

/1/ U.Kumpf, phys.stat.sol.(b) 52, 653 (1972)
/2/ H.Kiessig, U.Essmann, H.Teichler, W.Wiethaup, Phys.Lett. 51A, 333 (1975)
/3/ S.Foner, Rev.Sci.Instr. 30, 548 (1959)
/4/ F.C. von der Lage, H.A.Bethe, Phys.Rev. 71, 612 (1947)
/5/ S.J.Williamson, Phys.Rev. B2, 3545 (1970)

C-5 COMMENTS ON H_{c2} AT LOW TEMPERATURES

K.Takanaka

Department of Engineering Science, Tohoku University

Sendai 980, Japan

Two comments are made on the upper critical field at low temperatures: (1) on the upper critical field averaged over all field directions at zero temperature derived by Hohenberg and Werthamer and (2) on the temperature dependence of the upper critical field for Nb and V measured by Williamson at low temperatures.

(1) Hohenberg and Werthamer calculated the upper critical field appropriate for a polycrystalline sample averaged over all field directions at zero temperature /1/. Besides the inconsistency of the approximation /2/ the averaging procedure over the field directions is not correct, since they have not averaged the upper critical field for a single crystal over all field directions, although the errors in the final result may be quantitatively small.

The directional dependence of the upper critical field for a single crystal at $T=0$ is obtained in the following way. In the calculation of H_{c2}, we have always the term $VN(0)T \sum_{n=0}^{\infty} \frac{1}{|\omega_n|}$ which becomes $\ln(T_c/T)$, where V is the BCS parameter for the coupling strength, $N(0)$ the averaged density of states at the Fermi surface and $\omega_n = (2n+1)\pi T$. At $T=0$, the singular part of the temperature dependence $\ln(T_c/T)$ must be canceled by a term, which is constructed from the upper critical field H_{c2}, the temperature T and a component $v_\perp(\underline{p})$ of the Fermi velocity perpendicular to the applied magnetic field and with electron momentum $\underline{p}$. No other physical quantity appears in the calculation of H_{c2} (cf. Eq.(26) of ref./1/). The dimensionless parameter constructed from these three quantities is $(v_\perp^2(\underline{p})eH_{c2}^{(\alpha\beta\gamma)}/T^2)$, where α, β, γ are the directional cosines of the applied field with respect to the crystal axes and we have used the unit system $\hbar = k_B = c = 1$. To cancel the T dependence of $\ln(T_c/T)$ at $T = 0$, there must be a term of the functional form $-(1/2)\ln(Cv_\perp^2(\underline{p})eH_{c2}^{(\alpha\beta\gamma)}/T^2)$, where C is some constant. From these two terms, we have

$$<\ln(Cv_\perp^2(\underline{p})eH_{c2}^{(\alpha\beta\gamma)}/T_c^2)> = 0, \qquad (1)$$

where the brackets indicate the angular average over all the momenta at the Fermi surface. Thus, the upper critical field of the single crystal is obtained as

$$H_{c2}^{(\alpha\beta\gamma)} = (eT_c^2/C)\, e^{-<\ln(v_\perp^2(\underline{p}))>} \qquad (2)$$

This expression is identical with the expression for the upper critical field obtained in ref.3, if the constant C assumes the value $(2\pi^2/\gamma_0)$, where $\ln\gamma_0 = 0.577$.

To obtain the upper critical field appropriate for the polycrystalline sample, expression (2) must be averaged over all field directions, whereas Hohenberg and Werthamer /1/ performed the averaging procedure at the stage of Eq.(1).

(2) Williamson measured the upper critical field anisotropy of singlecrystalline Nb and V /4/ and analyzed the data in terms of the following expression

$$H_{c2}(t)/H_{c2}(0) = 1 + \eta\, t^2 \ln t \tag{3}$$

for reduced temperatures $t < 0.17$. The values η of Nb are found to be 1.69, 1.39 and 1.30 for the <111>, <110> and <100> crystallographic directions, respectively, while the value η for an isotropic superconductor is 0.634.

The theoretical expression for the upper critical field including the t^2-term is given by

$$H_{c2}(t)/H_{c2}(0) = 1 + 0.643\, t^2 \ln t - 0.854\, t^2. \tag{4}$$

The coefficient of the t^2-term has been calculated using the work of Maki and Tsuzuki /5/ for an isotropic superconductor. In the above temperature range the t^2-term gives quantitatively the same contribution as the $t^2\ln t$-term. Indeed, the behavior of $H_{c2}(t)/H_{c2}(0)$ with respect to $t^2\ln t$ is very similar to the behavior of a function $(1 + 2 \times 0.643\, t^2 \ln t)$. Thus, the discrepancy in η between experiment and theory can be removed, if the t^2-term and the anisotropy of the coefficients of the $t^2\ln t$- and t^2-terms /3/ are taken into account.

The upper critical field of Nb at very low temperatures $(t < 0.04)$ deviates from Eq.(3) and shows a t^2-dependence. This is due to the fact, that at such low temperatures the electron collision time τ due to impurities can not be neglected. Under the condition $\frac{1}{\tau} > t$ the upper critical field is given by

$$H_{c2}(t)/H_{c2}(0) = 1 + 0.643\, t^2 \ln(4.19/\tau) - 0.854\, t^2. \tag{5}$$

Interpolating the expressions (4) and (5) the upper critical field for small impurity concentrations is obtained:

$$H_{c2}(t)/H_{c2}(0) = 1 + 0.643\, t^2 \ln(t + 4.19/\tau) - 0.854\, t^2. \tag{6}$$

The detailed derivation of Eq.(5) and the extension to anisotropic superconductors will be given elsewhere.

The author thanks Dr.T.Nagashima, Professors T.Tsuzuki and T.Ohtsuka for useful discussions. He is grateful to Professor S.J.Williamson for suggesting the problem related to the second comment.

References

/1/ P.C.Hohenberg, N.R.Werthamer, Phys.Rev. 153, 493 (1967)
/2/ K.Takanaka, paper R-3, this volume, p.93
/3/ K.Takanaka, Progr.Theor.Phys. 46, 357 (1971)
/4/ S.J.Williamson, Phys.Rev. B2, 3545 (1970)
/5/ K.Maki, T.Tsuzuki, Phys.Rev. 139, 868 (1965).

C-6   THE INFLUENCE OF MAGNETIC ANISOTROPY ON THE
PROPERTIES OF NIOBIUM IN THE MIXED STATE

C.E.Gough

Department of Physics, University of Birmingham

Birmingham B15 2TT, U.K.

1. Introduction

For some time now at Birmingham we have been engaged in an extensive study of the thermodynamic and transport properties of niobium in the mixed state. Although our underlying aim has been to obtain information from these measurements about the properties of the elementary excitations in the mixed state, we recognised at an early stage the importance of taking full account of the anisotropic magnetic properties of niobium in any interpretation of our measurements [1]. Here we illustrate this by reference to our measurements of magnetization, heat capacity, thermal conductivity and ultrasonic attenuation, all of which are strongly influenced by anisotropy in the values of H_{c2}, κ_2 and B_0 (the field marking the transition between the intermediate mixed state and the fully developed Abrikosov phase).

2. Experimental Methods

In practice it is rather difficult to prepare niobium samples entirely free from magnetic hysteresis, since even the small amounts of damage introduced by normal handling can give rise to appreciable amounts of surface pinning of flux /2/. To indicate the severity of this problem, we have drawn in Fig.1 some typical minor hysteresis loops taken from magnetization measurements near H_{c2} for two typical samples prepared in the form of long cylinders with spark-planed ends for ultrasonic studies. The shape of these hysteresis loops suggests that both bulk and surface pinning are important in the first sample, whereas only surface effects contribute to the hysteresis in the second - the most reversible of all our ultrasonic samples. From such curves we see that not only are intrinsic properties like H_{c2} and the slope of the magnetization near H_{c2} anisotropic but so also is the extent of any hysteresis present.

There is no reason to believe that samples used by other authors to study the field dependence of the transport properties near H_{c2} are, in general, any more reversible than the examples shown. Unfortunately, many authors have been very uncritical in their interpretation of measurements in the presence of such hysteresis (some have even chosen to ignore it altogether, presenting measurements for either increasing or decreasing fields only). This has undoubtedly led to much of the confusion in the literature concerning the purity and field dependence of the mixed state properties of niobium and

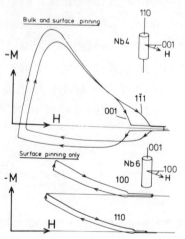

Fig.1:

Typical minor hysteresis loops for the magnetization near H_{c2} of samples prepared for ultrasonic studies.

vanadium near H_{c2} (see Purvis et al. /3/ for a summary of published ultrasonic measurements).

In view of the almost inevitable presence of hysteresis and the importance of clarifying the experimental situation, we would like to take this opportunity of urging all authors, when presenting measurements as a function of the externally swept field,
(a) to give examples of their measurements for both increasing and decreasing fields,
(b) to specify the field direction relative to the crystal symmetry directions, particularly when attempting to investigate differences in the transport properties measured parallel and perpendicular to the field direction,
(c) to publish magnetization curves for their samples in the form actually used in their experiments for increasing and decreasing H;
furthermore, our measurements suggest that authors should not assume (without supporting evidence)
(d) that measurements in either increasing or decreasing fields represent the ideal bulk behavior, or
(e) that the surface component of any hysteresis remains constant on passing through H_{c2}.

Fortunately, provided the amount of bulk pinning is not too large, rather reversible and hence reliable measurements can still be derived, if properties are measured directly as functions of flux within the sample /4/. In the next section we present some measurements of the anisotropy observed in the attenuation of sound near H_{c2}, which were derived in this way.

Recently we have developed a technique whereby ideally reversible measurements can be made /5/, even in the presence of bulk pinning, by deliberately introducing surface damage to inhibit any motion of flux through the surface /6/. When such a sample is allowed to cool through T_c in an external field, 100% flux trapping can be achieved; remarkably, this flux remains constant for any subsequent changes in external field below $\sim H_{c3}(T)$ and in temperature below $\sim T_{c3}(H)$. Any property can thus be measured as a function of temperature at constant and uniformly distributed flux throughout the mixed state. By varying the amount of flux trapped, the dependence of the property on B can be deduced. In the final section we present some measurements of the anisotropy of B_0 derived from ultrasonic attenuation measurements made in this way.

3. Anisotropy of H_{c2}, κ_2 and related properties

The anisotropy of H_{c2} in pure niobium is well known and will be discussed by other authors at this meeting /7/. Here we simply refer to Fig.2, where we have replotted some early measurements /1/ for the anisotropy of H_{c2}, and the anisotropy of the field dependence of the magnetization and ultrasonic attenuation observed in the same samples.

To obtain reliable measurements of the magnetization near H_{c2}, and hence values for κ_2, highly reversible samples must be used. Two such samples were prepared by Funnell /2/, who obtained values for κ_2 at 1.25 K of 2.33 ± 0.05 and 2.57 ± 0.03 along the <100>

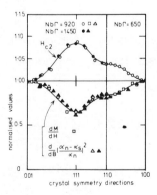

Fig.2: Anisotropy in the value of H_{c2} and in the slope of the magnetization and $\{(\alpha_n - \alpha_s)/\alpha_n\}^2$ near H_{c2} at 1.2 K.

and <110> cylindrical axis of the two single crystals, respectively. These values are consistent with the anisotropy observed for the slope of the magnetization in our earlier measurements shown in Fig.2, where we assumed the ideal magnetization lay mid-way between measurements in increasing and decreasing fields.

Below H_{c2} the transport properties are expected to be functions of Δ^2, the spatially averaged order parameter, which is proportional to the magnetization and hence to $(H_{c2} - H)$ and $(B_{c2} - B)$. Experimentally /4/, for fields not too close to H_{c2}, the attenuation in the mixed state α_s when plotted in the form $\{(\alpha_n - \alpha_s)/\alpha_n\}^2$ is proportional to $(B_{c2} - B)$ and therefore to the magnetization. From Fig.2 we see, that the observed field dependence in our ultrasonic measurements exhibits an almost identical anisotropy to that deduced for the magnetization. This supports the idea that real metal anisotropy can be largely taken into account in any microscopic theory, provided one takes values for Δ^2 derived from magnetization measurements along the specified field direction.

4. Anisotropy of B_o

The existence of the intermediate mixed state in niobium has been well established by decoration techniques /8/ and by neutron diffraction /9/. Information about the intermediate mixed state can also be derived from the sudden drop in magnetization observed at H_{c1} /10/. Here we consider how other properties are influenced by the formation of the intermediate mixed state, and will show how ultrasonic attenuation measurements can be used to derive values for B_o and its anisotropy.

We consider situations appropriate to our measurements in which we trap a constant amount of flux within the sample. For $B < B_o$ the fraction of the sample occupied by regions of Shubnikov phase (constant local induction B_0) is simply B/B_0. The influence of electronic excitations associated with these regions on any property will, in general, vary approximately linearly with B /6/. Above B_0, however, the properties of the electronic excitations will change as the flux line separation decreases, which results in a marked change in the field dependence at B_0, as observed in measurements of the heat capacity /11/, thermal conductivity /12/ and attenuation of sound /6/.

In Fig.3 we have plotted values of B_0 derived in this way from measurements of the attenuation of sound directly attributable to the electronic excitations in the regions of Shubnikov phase /6/. In the plane perpendicular to the <001> cylindrical axis of the crystal, along which the sound was propagating, B_0 rises to a maximum value of about 920 gauss with rather sharp minima a few degrees away from the <110> directions and a broader minimum as one approaches the <100> direction. Values obtained from similar measurements of the heat capacity /11/ and thermal conductivity /12/ along the <110> and <100> directions, respectively, and from neutron diffrac-

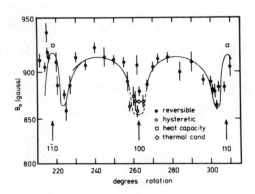

Fig.3: Anisotropy of B_0 at 1.8 K derived from ultrasonic attenuation measurements. Values obtained from independent measurements of the heat capacity /11/ and thermal conductivity /12/ have also been plotted.

tion /9/ along <110> are in good agreement with these measurements. However, a significantly lower value of B_0 ($\sim$ 750 gauss) has been derived from measurements of the heat capacity /11/ in a relatively impure crystal (resistance ratio $\sim$ 100) along the <100> direction. Away from the <100> direction there was no evidence for any hysteresis near the transition at B_0; however, at most angles close to <100> we observed significant hysteresis. This is illustrated by the measurements shown in Fig.4, where we have also plotted results for two other representative angles for comparison. For the measurements near <100> we have plotted the attenuation recorded both on heating from 1.2 K and on cooling from above 3 K (cycling the temperature inside these limits resulted in intermediate values of attenuation).

It seems rather likely that the hysteresis near the <100> direction is associated with a redistribution of flux lines below B_0, probably from a triangular lattice in the fully developed Abrikosov phase to a square lattice in the intermediate mixed state /8,13/. Under these conditions the transition at B_0 is no longer a continuous one, so that both superheating and supercooling are possible.

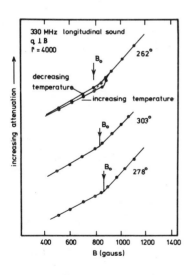

Fig.4:
Typical measurements of the field dependence of the attenuation at 1.8 K from which values of B_0 were derived. The attenuation scale is arbitrary and measurements at different angles have been displaced vertically for clarity. The angles given are the same as those plotted in Fig.3.

References

/1/ C.E.Gough, Solid State Commun. 6, 215 (1968)
/2/ I.R.Funnell, Ph.D. Thesis, University of Birmingham (1971)
/3/ M.K.Purvis, R.A.Johnson, A.R.Hoffman, J.Phys.Chem.Solids 35, 989 (1974)
/4/ E.M.Forgan, C.E.Gough, Phys.Lett. 26A, 602 (1968)
/5/ S.P.Farrant, C.E.Gough, Phys.Rev.Lett. 34, 943 (1975)
/6/ R.E.Jump, C.E.Gough, Proc.Int.Conf. Low Temperature Physics LT13 (Plenum, New York, 1974) vol.3, p.130
/7/ H.Teichler, paper R-1, p.7 ; T.Ohtsuka, paper R-2, p.27, this volume
/8/ U.Essmann, Phys.Lett. 41A, 477 (1972)
/9/ J.Schelten, H.Ullmaier, W.Schmatz, phys.stat.sol.(b) 48, 619 (1971)
/10/ U.Kumpf, phys.stat.sol.(b) 44, 829 (1971)
/11/ C.E.Gough, to be published
/12/ L.Lowe, W.F.Vinen, to be published
/13/ J.Schelten, G.Lippmann, H.Ullmaier, J.Low Temp.Phys. 14, 213 (1974).

C-7 ANISOTROPY OF THE STRESS DEPENDENCE OF CRITICAL
PARAMETERS IN UNIAXIAL SUPERCONDUCTORS

H.R.Ott

Laboratorium für Festkörperphysik, Eidgenössische Technische Hochschule Zürich, CH-8093 Zürich, Switzerland

A well known feature of experimental superconductivity is the change of the critical parameters (critical temperature T_c and critical field H_c) under the influence of external stress. Stress or pressure experiments are particularly useful in view of a comparison of experimental results with microscopic theories. Of special interest is the case of uniaxial stress applied on an anisotropic superconductor. Anisotropies are to be expected due to the non-cubic symmetry of the crystal lattice, which in turn influences the phonon and the electron spectrum of these axial materials.

We have chosen to study the non-transition metals tin, indium, gallium and zinc because the phonon and electron structures are well known at zero pressure. In order to study anisotropy effects the measurements have to be made on single crystal specimens. Experiments with real uniaxial stress on bulk single crystals are very difficult in these metals because of their low yield strengths.

In our case the stress derivatives of H_c and T_c were obtained from measurements of the volume or length change associated with the destruction of superconductivity in an external magnetic field and using the thermodynamic relation /1/

$$\frac{l_s - l_n}{l_s} \equiv \varepsilon_i = \frac{1}{4\pi} H_c \frac{\partial H_c}{\partial \sigma_i} + \frac{1}{8\pi} \kappa_i H_c^2 \qquad (1)$$

where κ_i is the uniaxial compressibility and σ_i is an uniaxial stress in the i-direction ($\sigma = -p$).

As an example we show in Fig.1 the temperature dependence of these relative length changes for tin single crystals oriented parallel and perpendicular to the tetragonal axis, respectively. From the temperature dependence of ε_i we obtain the temperature dependence of $\partial H_c/\partial \sigma_i$. The pressure derivative of T_c can now be determined in two ways using one of the following formulas

$$-\partial T_c/\partial \sigma_i = \frac{T_c}{H_o} \frac{1}{tf'} \{ (\partial H_c/\partial \sigma_i)_t - f(t)(\partial H_o/\partial \sigma_i) \} \qquad (2)$$

$$\partial T_c/\partial \sigma_i = - \frac{(\partial H_c/\partial \sigma_i)_{T_c}}{(\partial H_c/\partial T)_{T_c}} \qquad (3)$$

where $t = T/T_c$, $H_c = H_o f(t)$ and $f' = \partial f/\partial t$. H_o is the critical field at zero temperature. Equation (2) is based on a low temperature extrapolation of $\partial H_c/\partial \sigma_i$ and the existence of f(t). Equation (3) uses the high temperature extrapolation of $\partial H_c/\partial \sigma_i$ towards T_c. In our

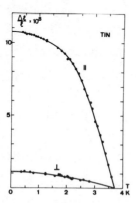

Fig.1:
Temperature dependence of the length changes of tin single crystals at the transition from the superconducting to the normal state in an external magnetic field,

∥ parallel to the tetragonal axis

⊥ perpendicular to the tetragonal axis.

case both equations give the same values for $\partial T_c/\partial\sigma$ within the possible error limits.

The analysis of our results is based on the McMillan equation for T_c /2,3/

$$T_c = \frac{\theta_D}{1.45} e^{-1/g} \qquad (4)$$

where

$$g = (\lambda-\mu^*-0.62\,\lambda\,\mu^*)/1.04(1+\lambda) \qquad (4')$$

θ_D is the Debye-temperature, λ is the electron-phonon interaction parameter and μ^* is the Coulomb potential of Morel and Anderson /4/. In a previous publication, Ott and Sorbello /5/ have shown how the stress dependence of T_c can be calculated using formula (4). Comparing theoretical and experimental values of $\partial T_c/\partial\sigma_i$ it turns out, that the anisotropy effects are mainly due to anisotropies of the phonon spectrum entering via the stress dependence of θ_D in our metals. The anisotropy of the stress dependence of the electronic structure can be important in metals, where the Fermi surface is close to Brillouin zone boundaries and the electronic density of states is very sensitive to changes of the lattice parameters. This seems to be the case in zinc and gallium, where the Fermi surface is known to have small bits and pieces.

Neglecting any pressure dependence of the Coulomb potential μ^* we can now calculate experimental values of the stress- or volume dependence of the electron-phonon interaction parameter λ using formula (4). In Table I we have listed these values determined from the experimental data of $\partial T_c/\partial\sigma_i$. Together with data on the pressure dependence of the electronic specific heat, which can be obtained from low temperature thermal expansion measurements, the stress dependence of the electronic density of states N(0) can now be calculated using the expression

Table I: Experimental values for the strain dependence of λ, calculated using equation (4). For more details see ref. /5/.

		Axis	$\partial \ln T_c/\partial \varepsilon_i$	$\partial \ln \lambda/\partial \varepsilon_i$
Indium	∥	tetrag.	6.5 ± 0.5	3.6
	⊥	tetrag.	5.3 ± 0.4	3.15
Tin	∥	tetrag.	12 ± 1	4.75
	⊥	tetrag.	4.6 ± 0.4	2.25
Zinc	∥	hex.	17.3 ± 1.5	3.25
	⊥	hex.	4.2 ± 0.5	1.1
Gallium		a	2 ± 1	1.85
		b	12.5 ± 1	11.6
		c	7.9 ± 1	7.35

$$C_p^{el} = \gamma T = \frac{1}{3}\pi^2 k_B N(0)(1+\lambda) T \qquad (5)$$

where k_B is the Boltzmann constant.

The basic ideas on the influence of small parts of the Fermi surface on the electron properties of metals are due to Lifshitz /6/. He predicted the occurence of electronic anomalies induced by actual topology changes (Lifshitz transitions) of the Fermi surface, when the Fermi surface is passing through a Brillouin zone boundary. Since the superconducting critical temperature is known to depend strongly on the electronic density of states, we expect anomalous changes of T_c associated with such a transition. Even more pronounced effects are very likely to occur in the behavior of $\partial T_c/\partial \sigma_i$.

To demonstrate this effect /7/ we have made the same measurements as described above on single crystals of α-phase In-Sn alloys. In this case the Lifshitz transition is achieved by increasing the Sn-content of these alloys from 0% to 12%. The increase of excess electrons in indium leads to the formation of new parts of the Fermi surface, the so-called α-arms /8/. This may clearly be seen in the stress dependence of T_c in these alloys shown in Fig.2.

Theoretical calculations for this system have been made by Hughes, Shepherd and Goulton /9/. For In-Sn alloys they predict the formation of the α-arms at a Sn-content of 6%, very close to the value of concentration where we observe an anomaly in $\partial T_c/\partial \sigma$. The change of anisotropy of $\partial T_c/\partial \sigma_i$ might be explained by the stress dependence of the various parts of the Fermi surface, as was shown in the case of gallium /10/. We note, that analogous to aluminium the newly formed α-arms should indeed have another sign for the stress dependence of their cross section than the β-arms /11/. This could be an explanation for the change of anisotropy of $\partial T_c/\partial \sigma_i$ observed in the experiments for In-Sn alloys. For indium the relevant measurements (stress dependence of the Fermi surface) have not yet been completed and similar conclusions as in ref. /10/ will have to await further investigations. More detailed information concerning this work has been given elsewhere /12,13/.

The author would like to thank Prof.M.J.Skove for his collaboration in part of this work.

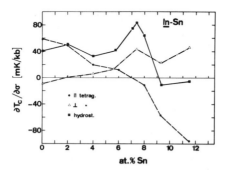

Fig.2: Stress dependence of the transition temperature as a function of alloy content. The hydrostatic pressure derivative is given by $\partial T_c/\partial \sigma = 2(\partial T_c/\partial \sigma_\perp) + (\partial T_c/\partial \sigma_\parallel) = -\partial T_c/\partial p$.

References

/1/ see e.g. D.Shoenberg, Superconductivity (Cambridge University Press, London 1965) p. 74
/2/ W.L.McMillan, Phys.Rev. 167, 331 (1968)
/3/ P.B.Allen, R.C.Dynes, Phys.Rev. B12, 905 (1975)
/4/ P.Morel, P.W.Anderson, Phys.Rev. 125, 1263 (1962)
/5/ H.R.Ott, R.S.Sorbello, J.Low Temp.Phys. 14, 73 (1974)
/6/ I.M.Lifshitz, Sov.Phys. JETP 11, 1130 (1960)
/7/ A very nice demonstration of this effect has been given by J.E.Schirber, Phys.Rev.Lett. 28, 1127 (1972)
/8/ We use the notation given by N.E.Ashcroft and W.E.Lawrence, Phys.Rev. 175, 938 (1968)
/9/ A.J.Hughes, J.P.G.Shepherd, D.F.Goulton, J.Phys. C3, 2461 (1970)
/10/ R.Griessen, H.Krugmann, H.R.Ott, Phys.Rev. B10, 1160 (1974)
/11/ R.Griessen, R.S.Sorbello, J.Low Temp.Phys. 16, 237 (1974)
/12/ H.R.Ott, ETH-Thesis 5077, 1973, unpublished
/13/ M.J.Skove, H.R.Ott, J.Low Temp.Phys. 25,   (1976)

R-3   MAGNETIZATION AND FLUX LINE LATTICE

IN ANISOTROPIC SUPERCONDUCTORS

K.Takanaka

Department of Engineering Science, Tohoku University

Sendai 980, Japan

Abstract

The anisotropies observed in magnetic properties such as the upper critical field, the magnetization and the correlation between the flux line lattice and the crystal lattice are considered in superconductors with uniaxial and cubic symmetry. The magnetization and the correlation are well explained by the effective mass model in uniaxial materials. As far as cubic materials are concerned, the correction of the order parameter to Abrikosov's solution turns out to be important. The upper critical field for polycrystalline samples is obtained by taking this correction into account. The term which is responsible for the correction of the order parameter plays an essential role in determining the anisotropy of magnetization and the correlation between the flux line lattice and the crystal lattice. The anisotropy of magnetization is in agreement with experimental observations, while the variety of correlations observed in Nb and Pb alloys cannot be explained by the extension of the Ginzburg-Landau equations, which includes only the lowest order term of non-locality.

1. Introduction

Tilley et al. /1/ discovered the anisotropy of the upper critical field of the cubic material Nb and tried to explain it using the Ginzburg-Landau equation with an anisotropic effective mass. In cubic systems, however, the effective mass is essentially isotropic. Hohenberg and Werthamer /2,3/ attributed this anisotropy to a nonlocal extension of the Ginzburg-Landau equation and to the anisotropy of the Fermi surface. Several investigations were made extending the work of Hohenberg and Werthamer to higher order terms in nonlocality /4-6/ or generalizing it to superconductors with gap anisotropy /7,8/.

If one includes corrections to Abrikosov's solution of the order parameter, which were not taken into account by Hohenberg and Werthamer, a variety of interesting results can be obtained; e.g., the upper critical field for polycrystalline samples is obtained by making this correction. Furthermore, the temperature dependence of the coefficients of the higher order cubic harmonics of the upper critical field can be evaluated. The nonlocal theory including an anisotropic Fermi surface can well explain the orientational dependence of the upper critical field as well as the temperature and purity dependence of the magnitude of anisotropy /9/.

The magnetization in anisotropic superconductors is in general not parallel to the applied magnetic field /10/. The pattern and the orientation of the flux line lattice are correlated to the Fermi surface anisotropy, i.e. to the crystal lattice /11-14/. Concerning the anisotropy of magnetization good agreement is obtained between experimental and theoretical results /4/ keeping only the lowest order term of nonlocality. However, the theory explains the observations on the correlation between flux line lattice and crystal lattice only partly /15,16/.

Theory of Magnetization and FLL   95

In this paper, we shall consider the magnetic properties of anisotropic superconductors with cubic symmetry near T_c. In addition to the effects mentioned above the existence of an anisotropy of the lower critical field is expected /17/. But since no experiment on the anisotropy of the lower critical field has been published so far, a discussion pertaining to this quantity will not be included in this report.

In section 2, we give qualitative arguments for the anisotropy of the magnetization and for the correlation between flux line lattice and crystal lattice in uniaxial systems. In section 3 the effect of anisotropy on the free energy, 3.1, the upper critical field, 3.2, the magnetization, 3.3, and the correlation between flux line lattice and crystal lattice, 3.4, in cubic systems will be discussed. In the following we use the unit system $\hbar = k_B = c = 1$.

2. Uniaxial Systems

In this section we consider layered materials such as $NbSe_2$ or the hexagonal elemental superconductor Tc. The mass in the layer plane is smaller than perpendicular to this plane for layered materials, while the mass along the c axis is smaller than in the basal plane for Tc.

The uniaxial system has a larger orientational dependence of the upper critical field H_{c2} than the cubic system. The ratio of $H_{c2\parallel}$ to $H_{c2\perp}$ depends on temperature, where $H_{c2\parallel}$ and $H_{c2\perp}$ denote the upper critical fields parallel and perpendicular to the layer plane for layered materials /18/ or to the c-axis for Tc /19/.

The effective mass model with gap anisotropy successfully explains the experimental results on the upper critical field of $NbSe_2$

and Tc /20/. Gap anisotropy of NbSe$_2$ is also consistent with an experiment on the far-infrared transmission spectrum /21/. Other theoretical models and experiments reported so far can be found in ref./8/.

In the following, we adopt the effective mass model to consider qualitative aspects of the anisotropy of magnetization and the correlation between the flux line lattice (FLL) and the crystal lattice (CL). Let us take technetium to serve as an illustration, because experimental results on the FLL are available. Theoretically, one can adopt the same treatment for layered materials by considering only the difference in the relative ratio of the effective masses.

For isotropic superconductors the magnetization is parallel to the external magnetic field and the flux line lattice is triangular. The mass difference in anisotropic superconductors leads to a distortion of the circular motion of electrons around the applied field and induces a component $M_\perp$ of the magnetization perpendicular to the magnetic field direction.

We choose the coordinate system (x,y,z), where the z axis is directed along the applied field, the y axis lies in the plane containing the c axis and the z axis, and the x axis is perpendicular to this plane.

By symmetry considerations, $M_\perp(\theta)$ has only a y component and shows twofold symmetry (θ is the angle between the c axis and the z axis, Figs. 1 and 2). The amplitude depends on the mass difference and decreases as $(H_{c2}-H_o)$ as the applied field H_o approaches the upper critical field H_{c2}.

As for the correlation between FLL and CL, we limit ourselves to qualitative arguments for the case, where the magnetic field is applied in the basal plane. By symmetry considerations the flux line

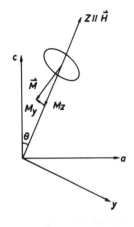

Fig.1:
Schematic representation of the electron orbit and the magnetization in technetium

lattice becomes an isosceles triangular lattice with one side in the basal plane or perpendicular to this plane. From the relation for the component of the current density $J_i \propto \langle v_i^2 \rangle \propto m_i^{-1}$, one obtains $|J_c| > |J_a|$, where the suffices a and c denote the components of $\underline{J}$ along the a and c axes, respectively. Using current conservation around each flux line, we expect a flux line lattice as shown in Fig.3.

Only one kind of flux line lattice (type a) is observed in experiment, which is in good agreement with detailed calculations /24, 25/.

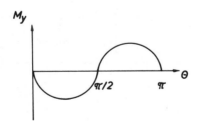

Fig.2: Angular dependence of the y-component of the magnetization

3. Cubic Systems

3.1 Free Energy

As a starting point we construct the free energy suitable to study the effect of anisotropy in cubic materials near the transition temperature T_c. The Ginzburg-Landau free energy is given as follows /26/,

$$F_{GL} = \frac{H_c^2 \lambda_T^3}{4\pi} \int d\underline{r} (h^2 - \frac{1}{2} + \frac{1}{2}(1 - |\psi|^2) + |\underline{q}\psi|^2), \qquad (1)$$

where H_c is the thermodynamic critical field, λ_T the penetration depth and all the lengths are measured in units of λ_T. We use the following notations for other quantities: the order parameter $\psi = \Delta/\Delta_{BCS}$ (Δ_{BCS} is the order parameter of a homogeneous superconductor), the magnetic field $\underline{h} = \underline{H}/(\sqrt{2} H_c)$, the vector potential $\underline{a} = \underline{A}/(\sqrt{2} H_c \lambda_T)$, $\underline{q} = i \underline{\nabla}/\kappa + \underline{a}$, where κ is the Ginzburg-Landau parameter.

Hohenberg and Werthamer attributed the anisotropy of the upper critical field in cubic materials to the anisotropy of the Fermi surface and to the nonlocality of the Ginzburg-Landau equations. The lowest order term of nonlocality is expressed by $<|(\underline{v}\,\underline{q})^2 \psi|^2>$ in the free energy, which corresponds to the η_{4c}-term in the

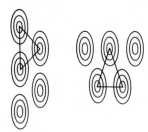

Fig.3: Schematic representation of the flux line lattice in technetium. Closed loops represent the current.

(a)   (b)

Neumann-Tewordt free energy /27/. Thus, the part of free energy giving the lowest order of anisotropy can be written as

$$F_{an} = \frac{H_c^2 \lambda_T^3}{4\pi} \int d\underline{r}\, \mu_4 <|(\underline{v}\,\underline{q})^2 \psi|^2>, \qquad (2)$$

where

$$\mu_4 = -\frac{9(1-t)S_{23}}{2<v^2>^2 S_{21}} < 0 \quad \text{and} \quad S_{ij} = \sum_{n=0}^{\infty} \frac{1}{(2n+1)^i\,(2n+1+\alpha)^j} \qquad (3)$$

The quantity α is the impurity parameter $(2\pi T_c \tau)^{-1}$. The total free energy is then given by

$$F = F_{GL} + F_{an}. \qquad (4)$$

Although this expression for the free energy does not contain all isotropic terms proportional to $(1-t)$, it is sufficient for the investigation of the lowest order effect of anisotropy on such quantities as the upper critical field H_{c2}, the lower critical field H_{c1}, the magnetization $M_\perp$ and the correlation between the flux line lattice and the crystal lattice.

To discuss the anisotropy of H_{c2}, H_{c1} and $M_\perp$ with the applied field in the (111) plane or the correlation between FLL and CL with the applied field along the <111> axis, it is necessary to add the next order term of nonlocality, $\mu_6 <|(\underline{v}\,\underline{q})^3 \psi|^2>$ to Eq.(2), where

$$\mu_6 = \frac{27}{4}\frac{(1-t)^2 S_{25}}{<v^2>^3 S_{21}} > 0 \qquad (5)$$

This is the first term, which shows threefold symmetry.

3.2 Upper Critical Field

The upper critical field H_{c2} is determined by the linearized Ginzburg-Landau equation, which is obtained from a variation of Eq.(4) with respect to ψ^*;

$$\psi = K(\underline{q})\,\psi \tag{6}$$

where

$$K(\underline{q}) = q^2 + \mu_4 \langle (\underline{v}\,\underline{q})^4 \rangle. \tag{7}$$

We take the direction of the applied field as the z axis and the vector potential as $\underline{a} = (0, hx, 0)$. The kernel $K(\underline{q})$ is separated into the diagonal part K_d and the off diagonal part K_{od} with respect to the eigenfunction of the operator q^2, namely,

$$K(\underline{q}) = K_d + K_{od} \tag{8}$$

where

$$K_d = \frac{2p+1}{2}\varepsilon + \frac{3}{16}(2p^2 + 2p + 1)\varepsilon^2 \mu_4 \langle v_\perp^4 \rangle, \tag{9}$$

and

$$K_{od} = \frac{\mu_4}{16} \{\langle v_+^4 \rangle q_+^4 + \langle v_\perp^2 v_+^2 \rangle \overline{q_+^3 q_-} + c\,c\,\}. \tag{10}$$

Here

$$q_\pm = q_x \mp iq_y,\quad v_\pm = v_x \pm iv_y,\quad v_\perp^2 = v_+ v_-,$$
$$\varepsilon = \frac{2h}{\kappa},\quad p = \frac{q_+ q_-}{\varepsilon}, \tag{11}$$

and the bar on $q_+^3 q_-$ indicates a summation over all permutations of q_- and q_+. We notice the following commutation relation for q_- and q_+

$$\{q_-, q_+\} = \varepsilon. \tag{12}$$

The Abrikosov solution ψ_0 of the operator q^2 is given by

$$\psi_0 = e^{-\kappa h x^2/2} \tag{13}$$

and satisfies the equation $q_-\psi_0 = 0$.

Expanding the order parameter ψ as

$$\psi(x) = \sum_{n=0}^{2} a_{2n} \frac{q_+^{2n}}{\sqrt{(2n)\varepsilon^{2n}}} \psi_0(x) \tag{14}$$

with $a_0 = 1$ and calculating the expectation value of Eq.(6), we determine the coefficients a_2 and a_4 by the variational principle under the condition $\partial h/\partial a_i = 0$ to obtain h_{c2}. The explicit expressions for a_2 and a_4 are as follows:

$$a_2 = \frac{3}{16\sqrt{2}} \mu_4 \varepsilon^2 <v_\perp^2 v_+^2> \quad \text{and} \quad a_4 = \frac{\sqrt{3}}{32\sqrt{2}} \mu_4 \varepsilon^2 <v_+^4> . \tag{15}$$

Substituting Eqs.(14) and (15) into Eq.(6) we obtain the upper critical field:

$$h_{c2} = \kappa\{H_1 - \frac{3}{4}\mu_4 <v_\perp^4> + \frac{3\mu_4^2}{16}[6<v_\perp^4>^2 + 6|<v_\perp^2 v_+^2>|^2 + |<v_+^4>|^2]\} , \tag{16}$$

where

$$<v_\perp^4> = \frac{4}{3}\{\frac{2}{5}<v^4>H_1 + \frac{3}{4}<v^4> - 5v_x^4>H_2\},$$

$$|<v_\perp^2 v_+^2>|^2 = <v^4> - 5v_x^4>^2(\frac{2}{105}H_1 - H_3 + H_4) , \tag{17}$$

$$|<v_+^4>|^2 = <v^4> - 5v_x^4>^2(\frac{8}{15}H_1 - 2H_1 - 10H_3 + H_4) .$$

Furthermore,

$$H_1 = 1 , \qquad H_2 = \alpha^2\beta^2 + \beta^2\gamma^2 + \gamma^2\alpha^2 - \frac{1}{5},$$

$$H_3 = \alpha^2\beta^2\gamma^2 - \frac{1}{105}, \qquad H_4 = \alpha^4\beta^4 + \beta^4\gamma^4 + \gamma^4\alpha^4 - \frac{1}{35}, \tag{18}$$

$$H_2^2 = \frac{4}{524} - \frac{2}{5}H_2 + 2H_3 + H_4 . \tag{18}$$

Here V_i is the i-th component of the Fermi velocity referring to the crystal axes (X,Y,Z) and α,β,γ are the direction cosines of the magnetic field with respect to the crystal axes.

If we add a contribution from the term $<(\underline{v}\,\underline{q})^6>$ in the kernel $K(\underline{q})$ to Eq.(16), we have the correct expression for H_{c2} up to $(1-t)^3$.

The following conclusions can be drawn from Eq.(16). First we notice, that the spatially averaged upper critical field H_{c2} is affected by the second order of the anisotropy parameter $<V^4 - 5V_x^4>$, i.e. the correction of the order parameter to Abrikosov's solution should be taken into account. The upper critical field $\overline{h_{c2}}$ at $T=0$ evaluated by Hohenberg and Werthamer is insufficient, because their expression for $\overline{h_{c2}}(0)$ was obtained by employing Abrikosov's solution of the order parameter. Thus, the discrepancy in $\overline{h_{c2}}(0)$ between the theory /28/ for an isotropic superconductor and experiment still remains /29,30/ an open question.

Generally, the second order terms of anisotropy cause an increase of the upper critical field and can be correlated to the positive curvature of $\overline{h_{c2}}$ at high temperatures.

Secondly, it is interesting to note that the coefficients of the third and fourth order cubic harmonics have the same temperature dependence near T_c. This fact has also been pointed out by Harms /5/, who calculated h_{c2} for a pure superconductor using a different method. Although Farrell et al. /31/ experimentally observed this fact in less pure Nb, it is desirable to measure the temperature dependence of the fourth order cubic harmonic of h_{c2} in other samples.

Theory of Magnetization and FLL   103

From experiments on the upper critical field one can obtain some information about the Fermi surface. Although experiments show $\langle V^4 - 5V_x^4 \rangle > 0$ and $\langle V^6 - 15V_x^4 V_y^2 + 30V_x^2 V_y^2 V_z^2 \rangle > 0$ for Nb, V and Pb-alloys, there has been no work giving quantitative estimates for these quantities.

In order to obtain the temperature dependence of the coefficients of the cubic harmonics in the entire temperature region in the case of small anisotropy, it is preferable to start from the theory of Helfand and Werthamer, who obtained the upper critical field including all orders of nonlocality. Small anisotropy can then be treated as a perturbation as was done by Teichler /7/.

If the pure superconductor shows anisotropy of the energy gap, the gap anisotropy is introduced as a weighting function in the density of states and the argument given above still holds, since the energy gap is a scalar function /2,7,8/.

3.3 Magnetization

The magnetization in anisotropic superconductors is not always parallel to the applied magnetic field. The component $M_\perp$ of the magnetization perpendicular to the magnetic field gives rise to a torque in the direction perpendicular to the applied field. We study $M_\perp$ near H_{c2} in this subsection.

To calculate $M_\perp$, it is necessary to use the second Ginzburg-Landau equation which is obtained by a variation of Eq.(4) with respect to the vector potential $\underline{a}$:

$$-2\,\mathrm{curl\,curl}\,\underline{a} = \underline{q}\psi\psi^+ + \mu_4 \langle \underline{v}\{(\underline{v}\,\underline{q})^3 \psi\psi^+ + (\underline{v}\,\underline{q})\psi(\underline{v}\,\underline{q})^2 \psi^+\}\rangle$$
$$+ c.c. \qquad (19)$$

At an external field h_0 just below the upper critical field, all the degenerate solutions of the kernel $K(q)$ must be superposed to construct the order parameter ψ;

$$\psi(\underline{r}) = C \sum_{n=0}^{\infty} a_{2n} \frac{q_+^{2n}}{\sqrt{(2n)!\,\varepsilon^{2n}}} \psi_0(\underline{r}) \tag{20}$$

and

$$\psi_0(\underline{r}) = \left(\frac{2x_{II}}{y_I}\right)^{\frac{1}{4}} \sum_{p=-\infty}^{\infty} \exp\left[\frac{2\pi}{y_I x_{II}}\left\{-\frac{(x+px_{II})^2}{2}\right.\right. \tag{21}$$

$$\left.\left. - ipx_{II}(y+\frac{p}{2}y_{II})\right\}\right]$$

where $\underline{r}_I = (0, y_I)$ and $\underline{r}_{II} = (x_{II}, y_{II})$ are the the lattice vectors of the primitive cell under the restriction of the flux quantization condition $2\pi/y_I x_{II} = \kappa h_0$. The coefficient C determines the magnitude of the order parameter and vanishes as $(1 - h_0/h_{c2})^{1/2}$ as $h_0 \to h_{c2}$.

We expand the vector potential and the magnetic field $(\underline{h}_0 + \Delta \underline{h})$ as

$$\underline{a} = C^2 \sum_{n,m=0}^{\infty} \underline{a}_{nm} q_+^n \psi_0 (q_+^m \psi_0)^+ , \tag{22}$$

and

$$\Delta \underline{h} = C^2 \sum_{n,m=0}^{\infty} \underline{h}_{nm} q_+^n \psi_0 (q_+^m \psi_0)^+ . \tag{23}$$

Substitution of Eq.(22) into Eq.(19) determines the coefficients $\underline{a}_{nm}$. Using the equation curl $\underline{a} = \underline{h}$, one obtains the elements $\underline{h}_{nm}$;

$$\underline{h}_{00}^{(\perp)} = -\frac{5\varepsilon}{16\kappa} \mu_4 \langle v_z v_\perp^2 \underline{v}_\perp \rangle$$

$$\underline{h}_{11}^{(\perp)} = \frac{\varepsilon}{16\kappa} \mu_4 \langle v_z v_\perp^2 \underline{v}_\perp \rangle \tag{24}$$

$$h_{00}^z = -\frac{1}{2\kappa}\left(1 + \frac{5\varepsilon\mu_4}{16} \langle v_\perp^4 \rangle\right)$$

$$h^z_{11} = -\frac{\mu_4}{2^5 \kappa} \langle v^4_\perp \rangle$$

$$h^z_{20} = \frac{\mu_4}{2^6 \kappa} \langle v^2_\perp v^2_+ \rangle \qquad (24)$$

$$h^z_{40} = \frac{\mu_4}{2^8 \kappa} \langle v^4_+ \rangle$$

where $\underline{0}^{(\perp)} = (0_x, 0_y)$.

Averaging $\Delta \underline{h}$ in the unit cell, we obtain for the magnetization $\underline{M}$

$$\underline{M}_\perp = \frac{3\varepsilon}{2^5 \kappa \pi} \mu_4 c^2 \langle v_z v^2_\perp \underline{v}_\perp \rangle \qquad (25)$$

and

$$M_z = -\frac{c^2}{8\pi\kappa}\left(1 + \frac{5\varepsilon}{16}\mu_4 \langle v^4_\perp \rangle\right). \qquad (26)$$

We consider the case, where the magnetic field is applied in the (100) plane. Without loss of generality, the x axis is chosen to be parallel to the <100> axis. If the z axis is rotated by an angle θ from the <001> axis, the angular average gives

$$\langle \underline{v}_\perp v^2_\perp v_z \rangle = \frac{\langle v^4 - 5v^4_x \rangle}{8} \sin 4\theta (0.1) \qquad (27)$$

Thus, when the applied field is in the (100) plane, the magnetization $M_\perp$ perpendicular to the field lies in the (100) plane and has fourfold symmetry. Its magnitude vanishes as $(h_{c2} - h_o)$ as $h_o \to h_{c2}$.

Recently, Hembach et al. /10/ made torque experiments on vanadium showing anisotropy of the magnetization in the mixed state. As a function of orientation the torque in the mixed state is observed to exhibit fourfold symmetry, while in the Meissner state twofold symmetry due to shape anisotropy was found. Furthermore, they determined the magnetization $M_\perp$ from the torque data and obtained the following results: (1) the orientation of the magnetization relative to the applied field direction displays fourfold symmetry with an extremum at $\pi/8$ with respect to the <100> axis; (2) the magnetization $M_\perp$ is found to be a function of the magnitude of the applied field and is a linear function of $(h_{c2} - h_o)$ near the upper critical field. If the experimental results on the upper critical field of vanadium are used, $<V^4 - 5V_x^4> > 0$ and Eqs.(25) and (27) are seen to be in agreement with the above observations.

3.4 Correlation between FLL and CL

The experimental methods to investigate the properties of the flux line lattice are the decoration technique /11,12/ and neutron diffraction /13,14/. Both methods have revealed, that the FLL is correlated with the crystal lattice in single crystalline superconductors.

Two sources of this correlation have been proposed: (1) anisotropy of the Fermi surface /15/ and (2) anisotropy of the magnetoelastic interaction /32/. The second proposal, however, was rejected by Roger, Kahn and Delrieu /16/, who made magnitude estimates of the two contributions. Therefore, only the influence of the anisotropic Fermi surface on the correlation is considered here.

Although Roger et al. made their investigations of the FLL near H_{c1}, we restrict ourselves to the case in which the external field is near to the upper critical field. We thus adopt arguments given by Abrikosov, base our calculation on the expression (4) for the free energy and discuss the correlation between FLL and CL, when the applied magnetic field is in the <100> direction.

Following Abrikosov's calculation of the free energy we have terms as $|\psi|^2$, $(\underline{h}_{c2} - \underline{h}_o + \Delta\underline{h})\Delta\underline{h}$ and $|\psi|^4$, where the bars denote spatial averages over the unit cell. The quantity $\overline{|\psi|^2}$ is a constant and depends neither on FLL nor on CL. If we keep the first order of anisotropy, the term $\overline{(h_{c2} - h_o + \Delta h)\Delta h}$ includes only the contribution from the z component which contains $|\psi_o|^2$, $\langle v_\perp^2 v_+^2 \rangle \overline{q_+^2 \psi_o \psi_o^+ |\psi_o|^2}$ and $\langle v_+^4 \rangle \overline{q_+^4 \psi_o \psi_o^+ |\psi_o|^2}$ and their complex conjugates. Thus, the stable FLL in anisotropic superconductors is determined by $\langle v_\perp^2 v_+^2 \rangle$, $\langle v_+^4 \rangle$ and β_{2n} where

$$\beta_{2n} = \frac{\overline{(q_+^{2n}\psi_o)\psi_o^+|\psi_o|^2}}{(\overline{|\psi_o|^2})^2 \varepsilon^{2n}} \qquad (28)$$

Therefore, the expression for β_o which determines the minimum of the free energy in isotropic superconductors is replaced by

$$f = \beta_o + \text{Re}\{C_2 \langle v_\perp^2 v_+^2 \rangle \beta_2 + C_4 \langle v_+^4 \rangle \beta_4\} \qquad (29)$$

Here, we have neglected the second order terms of anisotropy. The quantities $C_i (>0)$ depend on temperature, impurity concentration and the Ginzburg-Landau parameter. The free energy of the system depends on the pattern of the flux line lattice through β_{2n} and on the orientation with respect to the crystal lattice through $\langle v_\perp^2 v_+^2 \rangle$ and $\langle v_+^4 \rangle$. From the definition of β_{2n}, we have

$$\beta_o = \sqrt{z_1} \sum_{n,m=-\infty}^{\infty} K_{nm}$$

$$\beta_2 = \sqrt{z_1} \sum_{n,m=-\infty}^{\infty} K_{nm} \left(\frac{1}{2} - (n-m)^2 \pi z_1\right), \qquad (30)$$

$$\beta_4 = \sqrt{z_1} \sum_{n,m=-\infty}^{\infty} K_{nm} \left\{\frac{3}{4} - 3(n-m)^2 \pi z_1 + (n-m)^4 (\pi z_1)^2\right\},$$

where $z_1 = x_{II}/y_I$, $z_2 = y_{II}/y_I$ and

$$K_{nm} = \exp\{-(n^2+m^2)\pi z_1 + 2i\, nm\, \pi z_2\}.$$

By symmetry considerations we take $z_2 = 1/2$, which corresponds to an isosceles triangular lattice. The behavior of β_{2n} is shown in Fig.4. β_0, β_2 and β_4 assume the values 1.1596, 0.00 and 0.00, respectively, for a triangular lattice ($z_1 = \sqrt{3}/2$) and 1.1803, 0.00 and -1.41 for a square lattice ($z_1 = 1/2$).

We choose the z axis along the <001> axis and the x and y axes in the (001) plane. If the x axis is rotated from the <100> axis by an angle θ, the material parameters become

$$\langle v_\perp^2 v_+^2 \rangle = 0 \quad \text{and} \quad \langle v_+^4 \rangle = -\langle v^4 - 5v_x^4 \rangle e^{+4\theta i} \qquad (31)$$

The minimum of the free energy corresponds to the minimum of the function f with respect to θ, z_1 and δ, where

$$f(\theta, z_1, \delta) = \beta_0 - \delta \beta_4 \cos 4\theta \qquad (32)$$

and

$$\delta = 2 C_2 \langle v^4 - 5v_x^4 \rangle \qquad (33)$$

Rewriting Eq.(33) as

$$f(\theta, z_1 \delta) = \beta_0 - |\delta \beta_4| = f(z_1, \delta) \qquad (34)$$

we find a relation between z_1 and δ (Fig.5), for which $f(z_1,\delta)$ has

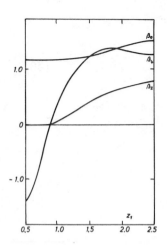

Fig.4: β_0, β_2 and β_4 versus z_1

a minimum. Thus, if δ is sufficiently large, lattice structures other than triangular ones may occur.

From Eqs.(32) and (33) the angle between the direction of the flux line lattice and the crystal axis is obtained as

$$\theta = \begin{cases} \dfrac{(2n+1)\pi}{4} & \text{for } \langle v^4 - 5v_x^4 \rangle > 0 \\ \dfrac{n\pi}{2} & \text{for } \langle v^4 - 5v_x^4 \rangle < 0 \end{cases} \qquad (35)$$

because $\beta_4(z_1)$ is negative in the range $1/2 < z_1 < \sqrt{3}/2$ and C_4 is positive. For Nb, V and Pb-alloys $\langle v^4 - 5v_x^4 \rangle$ is positive. Thus, the angle θ should be $(2n+1)\pi/4$ from Eq.(35).

The free energy, which involves only $\langle |(\underline{v}\,\underline{q})^2 \psi|^2 \rangle$ as the anisotropic term, allows us to discuss only the case, where the applied magnetic field is parallel to the <100> axis. If the magnetic field is parallel to the <111> axis, it is necessary to retain the term with threefold symmetry such as $\langle |(\underline{v}\,\underline{q})^3 \psi|^2 \rangle$ in the free energy expression. The term $\langle |(\underline{v}\,\underline{q})^2 \psi|^2 \rangle$ has no effect on the anisotropy, if the <111> direction is parallel to the applied field. The quantity $v_\perp^2 v_+^2$ has twofold symmetry and v_+^4 has fourfold symmetry about the z axis, but the <111> axis has threefold symmetry. Thus, $\langle v_\perp^2 v_+^2 \rangle$ and $\langle v_+^4 \rangle$ are zero, if the <111> direction is parallel to the field. For

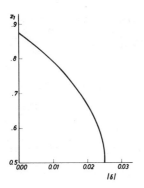

Fig.5:

$z_1 - |\delta|$ curve, which gives the minimum of $f(z_1, |\delta|)$

other directions parallel to the applied field, e.g. the <110> direction, it is necessary to include the term $<|(\underline{v}\,\underline{q})^3\psi|^2>$ in the free energy, which gives a contribution to the free energy of the same order as $<|(\underline{v}\,\underline{q})^2\psi|^2>$.

For arbitrary directions of the applied field such as the <431> direction, the condition $z_1 = 1/2$ is not satisfied and the flux line lattice is deformed from the isosceles triangular configuration.

Let us proceed to compare the theoretical and experimental results for the case $\underline{H}\|<001>$. The flux line lattice in Pb-alloys at low temperatures is a square lattice and its correlation with the crystal lattice is in good agreement with the above results. However, the theoretical results are partly in contradiction with the experimental observations on Nb.

There are several reasons to explain the discrepancies. The free energy Eq.(4), which we have employed, includes only the first order term of nonlocality. Experiments on the upper critical field suggest, that the nonlocal terms in Pb-alloys are small, while those in Nb are large. Thus, the higher order terms of nonlocality must be taken into account when calculating the free energy of Nb. For example, if we take the $<|(\underline{v}\,\underline{q})^3\psi|^2>$ term in the free energy, a $<v_-^2 v_+^4>$ term with fourfold symmetry appears.

Furthermore, there are other terms contributing to anisotropy. Werthamer has shown, that there is a term such as $(\nabla|\psi|)^2|\psi|^2$ in the free energy in the local limit /33/. This term is isotropic, but the next order or higher order terms of nonlocality may be anisotropic and may play some role in determining the stable flux line lattice.

Even if we take into account all of the possibilities discussed, there still remains a difficult problem. Two kinds of orientation of the square FLL relative to the CL have been observed in Nb and Pb-alloys experimentally. These two lattices are tilted with respect to each other by an angle of 30° at low temperatures. The decoration method also shows the coexistence of the Meissner state and the mixed state in such a situation. It is impossible to explain these flux line lattices within the framework of the present work. Therefore, many interesting problems in anisotropic superconductors remain to be solved.

The author thanks Dr.T.Nagashima and Prof.T.Ohtsuka for useful discussions.

References

/1/ D.R.Tilley, G.J. van Gurp, C.W.Berghout, Phys.Lett. 12, 305 (1964)
/2/ W.A.Reed, E.Fawcett, P.P.M.Meincke, P.C.Hohenberg, N.R.Werthamer, Proc. LT 10, Moscow, (Ed.M.P.Malkov), Viniti 1967, Vol. IIA, p. 368
/3/ P.C.Hohenberg, N.R.Werthamer, Phys.Rev. 153, 493 (1967)
/4/ K.Takanaka, T.Nagashima, Progr.Theor.Phys. 43, 18 (1970)
/5/ K.D.Harms, Z.Naturforsch. 25a, 1161 (1970)
/6/ T.Nagashima, Progr.Theor.Phys. 47, 37 (1972)
/7/ H.Teichler, phys.stat.sol.(b) 69, 501 (1975)
/8/ T.Nagashima, H.Fukuyama, to be submitted to J.Phys.Soc.Japan
/9/ T.Ohtsuka, paper R-2, this volume, p. 27
/10/ R.J.Hembach, F.K.Mullen, R.W.Genberg, Proc. LT 14 (H.Krusius, M.Vuorio, Eds.), North Holland 1975, Vol.2, p. 313
/11/ B.Obst, phys.stat.sol.(b) 45, 467 (1971)
/12/ U.Essmann, Phys.Lett. 41A, 477 (1972)
/13/ R.Kahn, G.Parette, Solid State Commun. 13, 1839 (1973)
/14/ J.Schelten, G.Lippmann, H.Ullmaier, J.Low Temp.Phys. 14, 213 (1974)
/15/ K.Takanaka, Progr.Theor.Phys. 46, 1301 (1971) and 50, 365 (1973)

/16/ M.Roger, R.Kahn, J.M.Delrieu, Phys.Lett. 50A, 291 (1974)
/17/ K.Takanaka, A.Hubert, Proc. LT 14 (H.Krusius, M.Vuorio, Eds.), North-Holland 1975, Vol.2, p. 309
/18/ N.Toyota, H.Nakatsuju, K.Noto, A.Hoshi, N.Kobayashi, Y.Muto, to be submitted to J.Low Temp.Phys.
/19/ G.Kostorz, L.L.Isaacs, R.L.Panosh, C.C.Koch, Phys.Rev.Lett. 27, 304 (1971)
/20/ K.Takanaka, phys.stat.sol.(b) 68, 632 (1975)
/21/ B.P.Clayman, Can.J.Phys. 50, 3193 (1972)
/22/ A.A.Abrikosov, Soviet Phys. JETP 9, 1364 (1959)
/23/ W.H.Kleiner, L.M.Roth, S.H.Autler, Phys.Rev. A133, 1226 (1964)
/24/ L.Dobrosavljevic, H.Raffy, phys.stat.sol.(b) 64, 229 (1974)
/25/ In Ref./20/ the flux line lattice of type b is missing. The angle θ is given by the conditions $n_1^{-1} x_2 / y_1 = 3/2$ and $y_1 = 2y_2$, and assumes the values 64° and 81° for Tc and $NbSe_2$, respectively.
/26/ V.L.Ginzburg, L.D.Landau, Zh.Eksp.i.Teor.Fiz. 20, 1064 (1950)
/27/ L.Neumann, L.Tewordt, Z.Physik 189, 55 (1966)
/28/ E.Helfand, N.R.Werthamer, Phys.Rev. 147, 288 (1966)
/29/ T.Ohtsuka, N.Takano, J.Phys.Soc.Japan 23, 983 (1967)
/30/ S.J.Williamson, Phys.Rev. B2, 3545 (1970)
/31/ D.E.Farrell, B.S.Chandrasekhar, S.Huang, Phys.Rev. 176, 562 (1968)
/32/ H.Ullmaier, R.Zeller, P.H.Dederichs, Phys.Lett. 44A, 331 (1973)
/33/ N.R.Werthamer, Phys.Rev. 132, 663 (1963).

R-4   MORPHOLOGY OF FLUX LINE LATTICES IN SINGLE
      CRYSTALLINE TYPE II SUPERCONDUCTORS

J.Schelten

Institut für Festkörperforschung, Kernforschungsanlage

D-517 Jülich, West Germany

Abstract

From experiments with the decoration technique and neutron diffraction by flux line lattices it was shown that in polycrystalline type II supercondcutors the flux lines form polycrystalline hexagonal vortex lattices as was predicted for isotropic superconductors by theory. In single crystalline superconductors which are anisotropic to some extent the following phenomena concerning lattice morphology were observed. (I) The vortex lattices are single crystalline. (II) There are correlations between crystallographic directions of the vortex and crystal lattice. (III) The crystal symmetry of the field axis influences the symmetry of the vortex lattice. These results were experimentally obtained in Nb, PbTl, PbBi, and Tc single crystals and the influence of various parameters on the vortex lattice morphology was investigated. These parameters are: the way of producing a vortex lattice, the thermodynamic variables as temperature and external field, the impurity parameter, the pinning strength, and dc and ac transport currents. Various experiments showed that the symmetry and the correlations of the vortex lattice are hardly affected by these parameters. However, the quality of the vortex lattice measured by the width of (10) rocking curves is easily influenced by almost all of these parameters and the pinning strength has the strongest effect.

Three quite different theoretical attempts were performed to connect the vortex lattice morphology to anisotropic properties of the superconductor, namely to the anisotropy of the energy gap, to the anisotropic magnetoelastic interaction between the vortex and the crystal lattice, and to the anisotropy of the Fermi surface. The results of these theoretical considerations will be compared with experimental data.

1. Introduction

The occurence of the Shubnikov phase also known as the mixed state was predicted by Abrikosov in his 1957 theory of superconductors with negative surface energy. This phase is characterized by a partial flux penetration, the flux being carried by quantized flux lines or vortices. Indirect proofs for the correctness of Abrikosov's concept were obtained from measurements of macroscopic properties of type II superconductors (e.g. magnetization curves). They showed good agreement with calculations based upon Abrikosov's theory. The first direct proof of the existence of flux lines in the mixed state was obtained by Cribier et al. (1964), who performed neutron diffraction experiments on superconducting niobium in the mixed state. In the first experiment of this kind a diffraction peak was found, which was consistent with the existence of a lattice formed by single quantized flux lines. From the following diffraction experiments Cribier et al. (1967) were able to demonstrate, that in polycrystalline niobium the flux lines formed a hexagonal rather than a square lattice. This result was obtained in the following way: The neutron beam of wave length λ_n is diffracted by the periodic arrangement of the flux lines. According to Bragg's law

$$2d_{hk} \sin(\tfrac{1}{2}\vartheta_{hk}) = \lambda_n \qquad (1)$$

maxima of the scattered neutron intensity occur at certain scattering angles ϑ_{hk}. In equation (1) d_{hk} is the flux line lattice spacing between (hk) lattice planes. For intensity reasons only the positions of (10) Bragg peaks were measured as a function of the flux density B in the mixed state. From the measured angles ϑ_{10} lattice spacings d_{10} were determined and compared with the lattice spacings calculated from the flux density B. It was assumed, that the magnetic flux of each vortex was equal to the flux quantum $\phi_0 = h/2e$ and that the vortices formed either a hexagonal or a square lattice. For these two cases, the lattice spacings are

$$d_{10}^{\triangle}(B) = \left(\frac{\sqrt{3}}{2}\frac{\phi_0}{B}\right)^{1/2} \quad \text{and} \quad d_{10}^{\square}(B) = \sqrt{\frac{\phi_0}{B}} \qquad (2)$$

respectively, and their relative difference amounts to 7%. The d_{10} values found from the Bragg peaks agreed well with $d_{10}^{\triangle}(B)$, while $d_{10}^{\square}(B)$ was outside the experimental uncertainty of d_{10}. Such a result was predicted from theoretical calculations of the free energy of the mixed state in isotropic type II superconductors showing that the free energy F_6 for a flux line lattice with sixfold symmetry was slightly smaller than the free energy F_4 for a lattice with fourfold symmetry. More recent numerical calculations based on the Ginzburg-Landau equations were performed in the entire mixed state by Brandt (1972). According to these calculations the relative difference $(F_4 - F_6)/F_6$ is at the most of the order of 1×10^{-3} and decreases to zero as B approaches either $\mu_0 H_{c2}$ or 0.

A decisive step towards a detailed investigation of the mixed state was provided by the development of the decoration technique by Essmann and Träuble (1967). This technique allows the observation of the positions of single vortices and led to the discovery of flux line lattice defects as dislocations, which are often arranged in the form of grain boundaries. Other defects, which were detected, are stacking faults, point defects as interstitials and vacancies and voids. The area F_c per flux line is directly obtained from a decoration picture and its magnification factor. By comparing F_c with $F_c(B) = \phi_0/B$ it was demonstrated again that the magnetic flux of each vortex is $1 \times \phi_0$. In addition, it was confirmed by this technique that in isotropic type II superconductors the vortices form more or less regular hexagonal lattices.

The properties of flux line lattices in <u>anisotropic</u> type II superconductors are strikingly different compared to the isotropic case. This statement results from the studies of the mixed state in single crystalline superconductors as Nb, V, Pb and Tc performed with the decoration technique and by neutron diffraction. The major morphological changes of the flux line lattices in anisotropic superconductors are: (1) The flux line lattices are single crystalline. (2) There are correlations between crystallographic directions of the flux line lattice and the crystal lattice. (3) The crystal symmetry of the field axis implies a certain symmetry of the flux line lattice. (4) The quality of the flux line lattice can be influenced by a large number of parameters.

In the following sections of this paper the experimental results on the vortex lattice morphology in anisotropic superconductors will be summarized and theoretical attempts discussed, which relate this vortex lattice property to other anisotropic but not necessarily superconducting properties of the crystals. It is hoped that this discussion will encourage theorists to improve and modify their theoretical calculations and to stimulate experimentalists to perform further experiments.

Morphology of Flux Line Lattices

2. Experimental

As already mentioned above, only two experimental techniques are known, which enable a determination of the morphology of flux line lattices. These two methods are based either on a decoration of flux lines or on neutron diffraction from vortex lattices. It is the purpose of this section to describe these two quite different techniques and to discuss their advantages and limitations.

Neutron scattering from the flux lines in the mixed state occurs due to the magnetic interaction between the nuclear magnetic moment of the neutrons with the microscopic field distribution in the mixed state. Because of the periodic arrangement of the vortices the diffracted intensity is peaked at certain scattering angles which are determined by the Bragg equation (1). In Fig.1 the centers of flux lines which form a hexagonal lattice are shown in a plane perpendicular to the external field H. This plane was also the scattering plane for all the diffraction experiments dealing with the morphology of the vortex lattice. The situation of neutron diffraction by (10) lattice planes is sketched in Fig.1. For this geometrical arrangement with H perpendicular to the scattering plane the Bragg condition for (10) lattice planes requires that the angle between the direction of the incoming neutron beam and the (10) lattice planes is $\vartheta_{10}/2$. Hence, rotating the single crystalline vortex lattice about the field axis and recording the diffracted neutron intensity as a function of the rotation angle φ an intensity pattern is obtained with maxima at certain φ values and zero intensity between these angles. For instance, after a rotation of 60° the neutron beam is diffracted by the (01) lattice planes and a rotation of 30° would be necessary to have the (11) lattice planes in Bragg condition. On the right hand side of Fig.1 the reciprocal vortex lattice is shown and the Bragg equation holds if

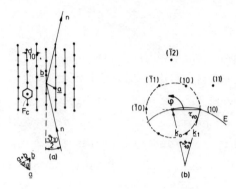

Fig.1: Vortex centers in a plane perpendicular to the external field. F_c area of a unit cell, $\underline{a}$ and $\underline{b}$ fundamental vectors of the vortex lattice and d_{10} spacing of (10) planes. The neutron beam of direction n is reflected by the (10) lattice planes. On the right, this Bragg reflection is represented by the neutron wave number vectors $\underline{k}_o$ and $\underline{k}_1$ and the reciprocal lattice vector $\underline{\tau}_{10}$. ϑ_{10} is the scattering angle, E is the Ewald sphere and φ the rotation angle of the reciprocal vortex lattice around an axis through the origin perpendicular to the drawing plane.

$$\underline{k}_o + \underline{\tau}_{hk} = \underline{k}_1$$

and

$$k_o = k_1 = 2\pi/\lambda_n \qquad (2)$$

In these equations $\underline{k}_o$ and $\underline{k}_1$ are the wave number vectors of the incoming and diffracted neutron beam and $\underline{\tau}_{hk}$ is a reciprocal lattice vector. Its length is $2\pi/d_{hk}$. Bragg reflection occurs if a reciprocal lattice spot is at the Ewald sphere (cf. Fig.1).

As an example, in Fig.2 the Bragg diffracted neutron intensity of a flux line lattice in single crystalline niobium with the field parallel to a <111> crystallographic direction is plotted versus the rotation angle φ. The most intense (10) maxima are separated by $60°$. Between such two peaks there is a less intense (11) maximum recorded at an scattering angle $\vartheta_{11} = \sqrt{3}\,\vartheta_{10}$. The height of these Bragg peaks is much smaller because the reflectivity decreases rapidly with increasing length of the reciprocal lattice vectors

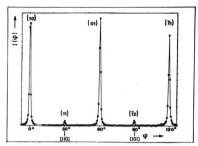

Fig.2: Neutron intensity I(φ) diffracted from the vortex lattice in single crystalline Nb with H parallel to the threefold <111> symmetry axis versus angle φ by which the vortex lattice is rotated around the field axis. The Bragg peaks of the vortex lattice are labeled by (hk); from Laue x-ray diffraction the crystallographic directions <110> and <101> marked on the φ axis were determined (T = 4.2 K, κ = 0.9, B = 800 G, Schelten et al. (1974)).

(Schelten et al. (1971) and (1972)). Fig.2 demonstrates that the flux line lattice is single crystalline and hexagonal. At the φ axis crystallographic directions of the niobium single crystal are also designated which were obtained from a x-ray Laue diffraction pattern. This information enables one to correlate the nearest neighbor directions $\underline{n}$ of the hexagonal vortex lattice with crystallographic directions of the crystal. In this case the $\underline{n}$'s were parallel to <110> directions.

If the hexagonal lattice were perfect, i.e. the (10) lattice planes were exactly parallel to each other, the peaks would have been much sharper. In this case their angular widths would be about 5', which corresponds to the instrumental resolution of this diffraction experiment. As can be seen in Fig.2 the angular width is about 60' and, therefore, on the average the lattice planes are tilted by this angle from their mean direction.

So far it has been demonstrated, that the four morphological properties of vortex lattices, which are generally observed in anisotropic superconductors as mentioned in the introduction, can all be determined in a neutron diffraction experiment. The major advantage of this technique is that the diffraction data are representative for the superconductor because of the following two reasons. Neutrons are scattered from vortices in the bulk of the specimen which typically has a volume of about 1 cm^3 and the scattering data are average values over a huge number of flux lines. This number is of the order of 10^{10} flux lines and the total length of investigated flux lines amounts to 100 000 km. In principle, the diffraction experiments can be performed in the entire temperature and field range of superconductors. It is, however, necessary that the field variation of the mixed state exceeds a certain value (about 50 G at a flux density of 1000 G in single crystalline superconductors) because otherwise the Bragg peaks can hardly be measured. Because of this sensitivity limit diffraction experiments were not successful in the vicinity of the transition temperature T_c, in the vicinity of the upper critical field H_{c2} and in high κ superconductors.

In the following the decoration technique will be described. A direct observation of the flux line lattice is obtained by decorating the points of exit of the flux lines at the specimen surface with small ferromagnetic particles. These particles are produced by a condensation process and are attracted by those points of the sample, where the vortex centers cross the top surface of the specimen. The specimen, e.g. a niobium rod, is half immersed in a bath of liquid helium. Three centimeters above the specimen a tungsten wire is installed from which iron is evaporated. The specimen is protected against heat radiation from the hot tungsten wire by a screening system of aluminium foils. The vapor pressure of helium is pumped down to about 0.8 torr corresponding to a temperature of the helium bath of 1.2 K. At this pressure the hot iron atoms shot from the tungsten wire have lost about 1 cm above the specimen surface almost all of their kinetic energy by collisions with the cold helium atoms and condense to small magnetic iron particles with a diameter of about 200 Å. The particles diffuse to the specimen surface and once they are attached to the surface they do not move any more. Close to the specimen surface the iron particles of magnetic moment μ are attracted by the force μ.grad H towards the centers of the flux lines and hence an accumulation of particles occurs at the places where the flux lines leave the specimen. After this decoration process the specimen is removed from the cryostat and the iron pattern stripped from the specimen surface by a carbon replica and observed in an electron microscope. For a more detailed description of this procedure reference is made to Träuble and Essmann (1966) and Essmann (1971).

As an example, in Fig.3 the flux line lattice in Pb - 1.6% Tl with the external field parallel to the fourfold <001> symmetry axis of the crystal is shown, Obst (1971). It is clearly demonstrated that the flux line lattice is single crystalline and that the flux lines or a square lattice which also has fourfold symmetry. The <100> and <010> directions of the lead crystal are determined by edge grooves. From this information the correlation between the nearest neighbor directions $\underline{n}$ of the flux lines and crystallographic directions of the crystal are established. In this case the $\underline{n}$'s were parallel to the <100> directions.

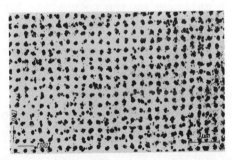

Fig.3: Vortex lattice in Pb-1.6% Tl with the external field H parallel to a fourfold <001> symmetry axis. ($\kappa_1 = 0.72$ at $T = 1.2$ K; H: $0 \to H_{c2} \to 365$ Oe.) (Reproduction of Fig.2 from Obst (1971).)

It is almost impossible to determine the quality of the flux line lattice quantitatively from decoration patterns, since for practical reasons only a very small fraction of the specimen surface can be investigated. For instance, Fig.3 contains about 400 vortices, which corresponds to a surface area of 7 µm x 4 µm. The decoration technique is superior to other experimental techniques when individual flux lines, lattice defects and selected portions of a vortex lattice are to be investigated. However, quantitative data of the vortex lattice morphology which are representative for the entire flux line ensemble are difficult to determine. Furthermore, up to now the decoration was always performed at a fixed temperature of 1.2 K and at relatively low flux densities $B \leq 1000$ G. The resolution of the decoration technique is roughly determined by the diameter of the black spots in the micrographs which is in the best cases 500 Å. This means that the vortex centers have to be separated by at least 2 times 500 Å corresponding to $B \approx 2000$ G, in order to be resolved clearly as single flux lines.

3. Vortex Lattices in Single Crystalline Superconductors

Morphological properties of vortex lattices determined in single crystalline PbTl, PbBi, Nb and Tc are listed in Table 1. All these results were obtained by the decoration technique or by neutron diffraction. If the external field is parallel to a fourfold symmetry axis of the crystals the vortices form a square lattice. In niobium such a square lattice was found only at low temperatures and at low flux densities. If the external field is parallel to a threefold symmetry axis the vortices form a hexagonal lattice. In the third case with H parallel a twofold symmetry axis always a distorted triangular vortex lattice was observed, in which two of three vectors a, b, and a+b (cf. Fig.1) had the same length and the third was either smaller or larger.

Table 1: Morphological properties of vortex lattices in single crystalline type II superconductors

Field direction	[001]	[011]	[1̄11]
Pb-1,6%Tl D$_1$	square, [010]/[100]	triangle 65°/65°/50°, [100]	triangle 60°/60°/60°, [211], [11̄2], [1̄2̄1]
Pb-Bi N$_1$	Pb-2% Bi, square, [010]/[100]	Pb-4% Bi, triangle 62°/62°/56°, [011̄]	
Nb N$_1$,N$_2$,N$_3$,N$_4$,D$_2$	T=1.5 K B≈B$_o$: square [1̄10]/[110], 30° rotation; T>1.5 K v B>B$_o$: triangle, 1̄00/[010], α=53.2°, η=63.4°	triangle 61,5°/61,5°/57°, [011̄]	triangle 60°/60°/60°, [110], [101], [01̄1]
	[112̄]	[205]	⊥[111]
Nb	triangle 57°/57°/66°, [11̄0]	triangle 55°/55°/70°	triangle ∡≈60°, [111]
	⊥[0001]		
Tc N$_1$	triangle 75°/52,5°/52,5°, [0001]		

D1 (Obst 1971), D2 (Essmann 1972); N1 (Schelten et al. 1974), N2 (Kahn et al. 1973), N3 (Thorel et al. 1973), N4 (Thorel et al. 1973)

In summary, if the external field is parallel to a four-, three- or twofold symmetry axis the vortex lattice has a four-, six- or twofold symmetry, respectively, apart from one exception (Table 1). The nearest neighbor direction $\underline{n}$ of the vortex lattice is parallel to low numbered crystallographic directions. A preferred direction for $\underline{n}$ is in Pb a <100> and in Nb a <110> crystallographic direction. Note that in the case H || <111> the projections of the <100> crystallographic directions onto the plane perpendicular to H define the nearest neighbor directions of the vortices in PbTl as well as the <211> directions.

The results of Table 1 were obtained in almost ideal typ II superconductors with fairly low κ-values. The question may be raised whether these results depend on (1) the kind of producing a vortex lattice, (2) the thermodynamic variables T and B, (3) the impurity parameter α, (4) the strength of flux pinning $P_v = B\,j_c$, (5) dc and ac transport currents with current densities below or above the critical current density j_c and (6) the shape of the specimen. A systematic investigation of the influence of all these parameters has not been performed because of the tremendous work which is involved with this multiparameter problem. However, in special cases the influence of one or the other parameter was studied and from these studies it can be concluded that only the pinning strength has a strong influence on the vortex lattice morphology. This will be demonstrated in the following subsections, where some experimental results referring to the points mentioned above are discussed. The results are important for a discussion of theories on the vortex lattice morphology, which will be given in section 4.

3.1 Four ways of producing a vortex lattice in a type II superconductor are sketched in the H-T diagram of Fig.4 showing the possibilities to reach a particular flux density B_1 and temperature T_1. (1): At T_1 the external field H is increased from zero to a

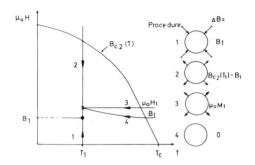

Fig.4: Four ways of producing a vortex lattice at temperature T_1 and of density B_1 are drawn in a H-T diagram schematically. The total flux density change ΔB which is connected with each procedure is tabulated and its sign is indicated by the arrows.

value $H_1 > B_1/\mu_0$ corresponding to the flux density B_1. In this procedure flux lines created at the specimen surface enter the bulk. The total flux change is $\Delta B = B_1$ as indicated in Fig.4. (2): At T_1 the external field is lowered from a value above H_{c2} to a value $H_1 > B_1/\mu_0$ corresponding to B_1. In this procedure flux lines created in the bulk of the specimen with a density $\mu_0 H_{c2}(T_1) > B_1$ are partly annihilated at the specimen surface. The total flux change is $\Delta B = \mu_0 H_{c2}(T_1) - B_1$. (3): At the field $H_1 > B_1/\mu_0$ the temperature is decreased from T_c to T_1. Flux lines are created in the bulk and the total flux change is M_1. (4): Here T and H are changed such as to keep the flux density constant. All four procedures were used once with a single crystalline niobium sample and H || <111>. The superconductor was almost ideal, had a κ value of 0.9 and an impurity parameter $\alpha = 0.1$. The results are as follows: No influence was observed on the symmetry of the vortex lattice nor on its correlation with the crystal. Only the degree of crystallinity (width of the rocking curve) improved slightly in going from procedure 1 to 2 to 3 to 4. It is likely that this kind of observed vortex lattice distortion is due to a small but non-zero effect of flux pinning. By changing the flux density as is done in procedures 1, 2 and 3, the maximum basic interaction force can be exerted to the flux lines, and the more the flux density is changed the more flux lines are exerted to this force. For a recent review on flux pinning see Ullmaier (1975). Furthermore it should be mentioned that the maximum of the basic interaction force is usually monotonically increasing with decreasing flux density (Lippmann et al. (1975), Antesberger and Ullmaier (1974), Freyhardt (1971)). Following these arguments it becomes qualitatively clear, that the broadest rocking curves were observed when the vortex lattice was produced according to 1, since this procedure involves a large flux density change $\Delta B = B_1$ at low flux densities $0 \leq B \leq B_1$. On the other hand the sharpest rocking curves were measured with procedure 4 involving no flux change, since the pinning forces are kept minimum.

3.2 The influence of temperature and of flux density has been investigated frequently (Schelten et al. (1974), Weber et al. (1974), Thorel et al. (1973)). Generally, only the quality of the vortex lattice was influenced by T and B. The only exception was found in niobium with H $\parallel$ <100> at low T and B. In this case two square lattices were detected by both techniques, while at higher T or B the flux lines form two distorted triangular lattices.

3.3 A very important result for the further discussion is that in dirty as well as in pure niobium single crystals the vortices form single crystalline vortex lattices (Weber et al. (1973) and (1974)). By dissolving nitrogen in niobium single crystals the impurity parameter $\alpha = 0.882$ (ξ_0/ℓ) was varied between 0.03 and 2. ξ_0 is the coherence length of the pure superconductor and ℓ is the electron mean free path of the dirty superconductor. It is found that the impurity parameter α did not influence the sixfold symmetry of the vortex lattice when H was parallel to a <111> crystallographic direction nor the distortion of the vortex lattice with H parallel to the twofold axes <110> and <230>. Merely the quality of the vortex lattice was different for the various specimens. The increase of the widths of the rocking curves with increasing impurity is probably again caused by different pinning strengths. Flux pinning was indicated by the hysteresis of the magnetization curves. Furthermore, it was demonstrated by transmission electron microscopy (TEM) that non-superconducting Nb_2N precipitates were present which are known as strong pinning centers.

3.4 In a neutron diffraction experiment made to determine the basic interaction forces for flux pinning, niobium single crystals with Nb_2N precipitates of number densities between 10^{11} and 10^{12} cm^{-3} were investigated (Lippmann et al. (1976)). The diffraction experiments demonstrated that sufficiently strong pinning inhibits the formation of a single crystalline vortex lattice. The external field was parallel to a <111> crystallographic direction. The vortex lattice was produced by cooling the specimen from a temperature above T_c to 4.2 K in a constant external field H_1 (procedure 3). Thereby a homogeneous flux density $B = \mu_0 H_1$ is obtained in the bulk of the specimen, while in the vicinity of the specimen surface the flux density drops to $B = \mu_0 H_1 - M_1$, where M_1 denotes the magnetization of the same superconductor without flux pinning. The width η of the rocking curves versus the critical current density j_c at $B/B_{c2} = 1/2$ is shown in Fig.5. At critical current densities exceeding 1.5×10^8 A/m^2 the vortex lattice is polycrystalline while at j_c's below 1×10^8 A/m^2 the single crystalline vortex lattice had rocking curve widths smaller than $10°$. At $j_c \approx 0$, i.e. in almost ideal niobium single crystals, η values smaller than $1°$ were measured.

3.5 The influence of dc and ac transport currents on the quality of vortex lattices has been investigated by neutron diffraction in Saclay (Simon and Thorel (1971) and Thorel et al. (1972)). The results of these measurements may be summarized as follows: In the zero voltage region ($I < I_c$), i.e. with transport currents I less than the critical current I_c, the width of the rocking curve was unchanged. In the non-linear region with $I \geq I_c$ the rocking curves were broadened, while in the linear flux flow region ($I > 2I_c$) the

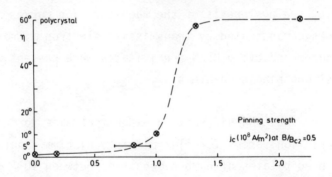

Fig.5: The width η of (10) rocking curves of vortex lattices in single crystalline Nb with H parallel to the threefold <111> symmetry axis versus the pinning strength characterized by the critical current density j_C at the reduced flux density of 0.5. Flux pinning was caused by Nb_2N precipitates and varied by the number density of precipitates.

narrowest rocking curves were measured. In this region the widths of the rocking curves were typically 10'. In these experiments the current was applied perpendicular to the external field and the neutron beam. Thus, the driving force density $\underline{P}_D = \underline{B} \times \underline{j}$ exerted to the flux lines was in the direction of the neutron beam. If $|\underline{P}_D|$ exceeds the pinning force density $\underline{P}_V = \underline{B} \times \underline{j}_c$, the flux lines perform a viscous motion. At high transport current densities $j > j_c$, i.e. in the linear flux flow region, there is a tendency to align the lattice planes by the driving force and the dragging force. Both have a uniform direction of opposite sign within the specimen, in contrast to the pinning forces which act on the flux lines in all directions, but its average is also opposite to the driving force.

3.6 There is little experimental evidence that the shape of the specimen influences the vortex lattice morphology. Thorel and Kahn (1973) have found that the nearest neighbor direction $\underline{n}$ deviates slightly from a <110> crystallographic direction in niobium depending on the shape and orientation of the cross section of their rodlike crystals. Specimens with rectangular and trapezoidal cross sections were investigated.

It can be concluded, that the symmetry of the vortex lattice and the correlations between nearest neighbor directions in the vortex lattice and crystallographic directions of the crystal lattice are hardly influenced. However, the quality of the vortex lattice depends on almost all of the parameters mentioned above, the pinning force playing the dominant role.

4. Theories of the Vortex Lattice Morphology

In single crystalline superconductors the nearest neighbor directions $\underline{n}$ of the vortex lattices are correlated with crystallographic directions and the shape of the unit cell of the vortex lattice depends on the crystallographic direction parallel to the field. These two experimental facts imply, that the free energy of the mixed state in single crystalline superconductors is a function of three more variables in addition to those in isotropic superconductors. These variables are, e.g., the angle φ between $\underline{n}$ and a crystallographic direction and the angles α and β of the triangle with the sides $\underline{a}$, $\underline{b}$ and $\underline{a}+\underline{b}$ as shown in Fig.1. With such an expression for the free energy $F(\varphi,\alpha,\beta)$ the vortex lattice morphology would be determined by minimizing F with respect to the variables φ, α and β. Three theoretical attempts have been made so far to estimate the anisotropy of the free energy by assuming that it is caused solely (a) by the anisotropy of the gap parameter Δ, (b) by an anisotropic magnetoelastic interaction between the vortex lattice and the crystal and (c) by the anisotropy of the Fermi surface. A more general theoretical investigation including the anisotropy of the coupling constant is discussed by Teichler (1974).

In this section I will compare experimental results with theoretical predictions. This comparison will show, that none of the theoretical attempts is able to describe details of the experimental results.

(a) Obst (1971) assumed that the interaction energy between two flux lines in a low κ superconductor had the same functional relations as were derived for superconductors with $\kappa \gg 1$. Under this assumption it was shown, that for a fixed vortex distance d the interaction energy is direction-dependent and has a maximum and a minimum in those directions of d, where the energy gap is minimum and maximum, respectively. Qualitatively, the correlations of the vortex lattice in PbTl crystals could be explained. However, there are two experimental facts which are seriously in conflict with this treatment:

(I) In pure niobium the anisotropy of the energy gap is known from the tunneling experiments of MacVicar and Rose (1968). Its angular dependence in a (111), (112) and (110) plane is plotted as a polar diagram in Fig.6. The diagram contains also the nearest neighbor directions of the vortex lattices, which were determined by applying the external field perpendicular to these planes. Fig.6b shows that the nearest neighbor directions coincide with directions of minima of the energy gap which is in conflict with Obst's conclusion. Furthermore, in Fig.6a the energy gap is direction-independent but the correlations do not disappear. Finally, in Fig.6c it is demonstrated, that the nearest neighbor directions coincide with directions of both maxima and minima of the energy gap.

(II) In tunneling experiments on niobium crystals with different purity, a gap anisotropy was measured only in relatively pure samples, i.e. for residual resistance ratios RRR $\geq$ 65. This result is consistent with the theory (Anderson (1959)), that the anisotropy is smeared out by electron scattering from impurities. Therefore, one should expect, that the correlations of the vortex lattice disappear in dirty superconductors, if the gap anisotropy alone were responsible. This is, however, in contrast to the experimental result discussed in the previous section and showing that in dirty Nb

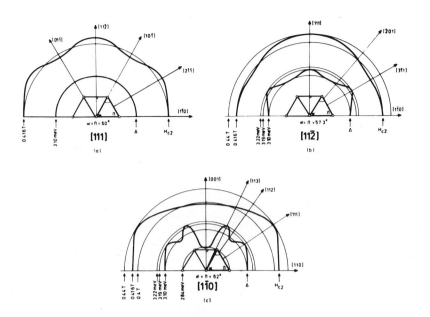

Fig.6: In polar diagrams the direction dependence of the upper critical field H_{c2} and the energy gap Δ for Nb at T = 0 are plotted for the (111), (112) and (110) planes. The nearest neighbor directions of the flux line lattice with H perpendicular to these planes are included. Data for H_{c2}, Δ and the vortex lattice from Williamson (1970), MacVicar and Rose (1968), and Kahn and Parette (1973), respectively. (Reproduction of Fig.2 from Kahn and Parette (1973).)

specimens with RRR values less than 10 correlations did exist. Hence, it may be concluded that correlations can only be related to the gap anisotropy in pure superconductors. Such a relation is, however, much more complicated than the one proposed by Obst.

(b) A quite different theoretical attempt was performed by Ullmaier et al. (1973). The authors considered a coupling of the vortex lattice and the crystal lattice through a magnetoelastic interaction which depends on the anisotropic elastic properties of the crystal. Since the elastic constants are almost independent of the impurity of the superconductor, this mechanism is not restricted to pure superconductors. For the direction dependence of the magnetoelastic interaction energy the following expression was calculated:

$$E_e = \left(\frac{\Delta V}{V}\right)^2 \left(\frac{c_{11}+2c_{12}}{3}\right)^2 \sum_{hk} \gamma(q_{hk})|F_{hk}|^2 \qquad (3)$$

$$\gamma^{-1}(\underline{q}) = c_{12} + c_{44} + \left(\sum_{i=1}^{3} \frac{q_i^2}{c_{44}+dq_i^2}\right)^{-1} \qquad (4)$$

and

$$d = c_{11} - c_{12} - 2c_{44} \qquad (5)$$

In these equations ΔV is the difference between the specific volumes of the metal in the normalconducting vortex cores (V) and in their superconducting surrounding (V - ΔV). The c_{ij} are the elastic constants of the crystal with the anisotropy factor d. The argument of γ is a reciprocal lattice vector of the vortex lattice normalized to unity and with components in direction of the cubic axes q_1, q_2 and q_3. Note, that the magnetoelastic energy is direction-dependent only if $d \neq 0$, since with $d = 0$ $\gamma(\underline{q})$ is a constant. F_{hk} is the form factor of the strain field within a unit cell of the vortex lattice and normalized to 1 at $\tau_{hk} = 0$. If one assumes, that the local strain is proportional to the microscopic field h(r), $|F_{hk}|^2$ is identical with the form factor of h(r) and in this case $|F_{hk}|^2$ is strongly dependent on the length of the reciprocal lattice vectors τ_{hk}. For the two cases that the external field is parallel to a fourfold and a threefold axis, respectively, the symmetry and the correlation of the vortex lattice were determined by minimizing E_e of equation (3).

The results are in reasonable agreement with the experimental data of Table 1 (Ullmaier et al. (1973)). In the following the magnitude of E_e will be estimated and compared with the interaction energy between vortices and pinning centers. In niobium with $\Delta V/V = 3 \times 10^{-7}$ and $\underline{H}$ parallel to a <100> axis the difference between the maximum and the minimum of E_e is $\Delta E_e = 6 \times 10^{-3}$ Ws/m^3. This energy per unit volume can be gained, if the vortices which have formed a square lattice with the nearest neighbor directions in <100> directions rearrange to another square lattice with n parallel to <110>. For this 45° rotation about the field axis an average displacement per vortex of the order of the lattice parameter a is necessary. In a superconductor with flux pinning mechanical work has to be done to perform the displacements and for randomly distributed pinning centers this energy contribution can be estimated to be

$$E_p = j_c B a$$

Using typical values for an almost ideal niobium crystal (j_c = 0.5×10^6 A/m² at B = $0.5 \mu_0 H_{c2}$) E_p = 8.5×10^{-3} Ws/m³ is obtained which is close to the value of ΔE_e given above. As a consequence of this comparison we would expect that the correlations disappear in irreversible superconductors, if the magnetoelastic interaction alone were responsible for the correlations. However, correlations have been measured in niobium single crystals, where the pinning energy was 100 times larger (cf. Fig.5).

(c) Another theoretical treatment of the correlations of vortex lattices is due to Takanaka, who calculated the free energy of the mixed state with an anisotropic Fermi surface. This treatment starts from the Ginzburg-Landau equations and contains higher order terms with respect to the spatial derivatives of the magnetic field and is restricted to the case of pure superconductors and temperatures near the transition temperature. Near the upper critical field H_{c2} the free energy of the mixed state is given by (Takanaka (1973b))

$$F = \frac{1}{2\mu_0} (B^2 - \frac{(B - \mu_0 H_{c2})^2}{(2\kappa_2'^2 - 1)\beta_0 + n'})$$

where κ_2' and n' are complicated functions of higher moments of the Fermi velocity, of the direction cosines of the external field with respect to the cubic axes, and of the shape and orientation of the unit cell. The effort to obtain quantitative information about the symmetry and correlations of the vortex lattices by minimizing F faces the problem, that the anisotropy of the Fermi surface is not known well enough for niobium, lead or technetium. Nevertheless, in some cases quantitative results could be obtained using experimental values for the direction-dependence of H_{c2}. The theoretical predictions obtained in this way are in reasonable agreement with the experimental results of Table 1. It should be mentioned that the directions in which maxima or minima of H_{c2} occur are not directly related to the nearest neighbor directions of the vortex lattices. This is demonstrated in Fig.6, where the direction-dependence of H_{c2} is plotted in a polar diagram. In Fig.6a, the directions of H_{c2} minima coincide with the nearest neighbor directions, while this is not the case in Fig.6b and c.

In order to establish the anisotropy of the Fermi surface as being responsible for the correlations in irreversible superconductors, the total change of F resulting from a rotation of a vortex lattice about the field axis must exceed a certain limit. The arguments for this requirement are the same as given above in connection with the magnetoelastic interaction. From expressions and data presented by Takanaka (1971) I have, therefore, estimated this total change of F for the case of Nb with H parallel to a <100> crystallographic direction, and obtained:

$$\Delta F = 0.009(\mu_0 H_{c2} - B)^2 \frac{1}{2\mu_0}$$

In order to compare ΔF with the values calculated above, $B = 0.5\,\mu_0 H_{c2}$ will be inserted, although I am aware of the fact that Takanaka's theory is valid only in the vicinity of H_{c2}. Using H_{c2} of pure niobium one obtains $\Delta F = 70$ Ws/m^3 which is four orders of magnitude larger than the corresponding variation of the magnetoelastic interaction energy. Hence, it may be concluded that the anisotropy of the Fermi surface could be responsible for the correlations observed in superconductors with strong flux pinning.

Takanaka (1975) has extended his theory to the case of superconductors with unaxial symmetry, e.g. layered superconductors, and applied it to technetium which crystallizes in a hexagonal lattice. Again, quantitative results could be obtained using the direction dependence of H_{c2} in technetium, which was measured by Kostorz et al. (1971). The theoretical calculations yielded a distorted triangular flux line lattice with an isosceles triangle, whose base angles are $54°$. This result agrees well with the neutron diffraction result for Tc in Table 1.

In this section the three existing theoretical attempts to explain the correlations between vortex and crystal lattices were reviewed. It was demonstrated that the predictions, which were derived from the anisotropy of the Fermi surface, agreed well with experimental results. Despite this success it is still an open question whether the anisotropy of the Fermi surface alone is responsible for the correlations. To answer such a question we need more quantitative predictions from theoretical calculations and experimental data to compare with.

I am grateful to Dr.G.Lippmann, with whom I performed the neutron diffraction experiments, and to Dr.H.Ullmaier for a fruitful co-operation over several years. Furthermore, I wish to thank Prof. W.Schmatz for his continuous interest in my research and for innumerable, valuable discussions.

References

A.A.Abrikosov, Soviet Phys. JETP 5, 1174 (1957)
G.Antesberger, H.Ullmaier, Phil.Mag. 29, 1101 (1974)
E.H.Brandt, phys.stat.sol.(b) 51, 345 (1972)
D.Cribier, B.Jacrot, L.M.Rao, B.Farnoux, Phys.Lett. 9, 106 (1964)
U.Essmann, H.Träuble, Phys.Lett. A24, 526 (1967)
U.Essmann, Physica 55, 83 (1971)
U.Essmann, Phys.Lett. 41A, 474 (1972)
D.E.Farrell, B.S.Chandrasekhar, S.Huang, Phys.Rev. 176, 562 (1968)
H.C.Freyhardt, Phil.Mag. 23, 345 (1971)
H.C.Freyhardt, Phil.Mag. 23, 369 (1971)
R.Kahn, G.Parette, Solid State Commun. 13, 1839 (1973)
G.Kostorz, L.L.Isaacs, R.L.Panosh, C.C.Koch, Phys.Rev.Lett. 27, 304 (1971)
G.Lippmann, J.Schelten, W.Schmatz, Phil.Mag. 33, 475 (1976)
M.L.A.MacVicar, R.M.Rose, J.Appl.Phys. 39, 1721 (1968)
B.Obst, phys.stat.sol.(b) 45, 467 (1971)
J.Schelten, H.Ullmaier, W.Schmatz, phys.stat.sol.(b) 48, 619 (1971)
J.Schelten, H.Ullmaier, G.Lippmann, Z.Physik 253, 219 (1972)
J.Schelten, G.Lippmann, H.Ullmaier, J.Low Temp.Phys. 14, 213 (1974)
K.Takanaka, T.Nagashima, Progr.Theor.Phys. 43, 18 (1970)
K.Takanaka, Progr.Theor.Phys. 46, 357 (1971)
K.Takanaka, Progr.Theor.Phys. 46, 1301 (1971a)
K.Takanaka, Progr.Theor.Phys. 49, 64 (1973)
K.Takanaka, phys.stat.sol.(b) 68, 623 (1975)
H.Teichler, Phil.Mag. 30, 1209 (1974)
H.Teichler, Phil.Mag. 31, 775 (1975)
P.Thorel, R.Kahn, Y.Simon, D.Cribier, J.Phys. 34, 447 (1973)
P.Thorel, R.Kahn, ANL-Report 8054 (1973), Argonne Nat.Lab., p. 50
H.Träuble, U.Essmann, phys.stat.sol. 18, 813 (1966)
H.Ullmaier, R.Zeller, P.H.Dederichs, Phys.Lett. 44A, 331 (1973)
H.Ullmaier, Springer Tracts in Modern Physics 76, edited by

G.Höhler, Springer Verlag (1975)

H.W.Weber, J.Schelten, G.Lippmann, phys.stat.sol.(b) 57, 515 (1973)

H.W.Weber, J.Schelten, G.Lippmann, J.Low Temp.Phys. 16, 367 (1974)

S.J.Williamson, Phys.Rev. B2, 3545 (1970)

C-8 CORRELATIONS BETWEEN FLUX LINE LATTICE
AND CRYSTAL LATTICE

B.Obst

Institut für Experimentelle Kernphysik der Universität
und des Kernforschungszentrums, D-75 Karlsruhe, W-Germany

1. Introduction

The original theory of superconductivity /1/ considered only homogeneous systems, in which the pair potential Δ was independent of space and time. In choosing an effective phonon-induced electron-electron interaction V the theory ignored a detailed energy dependence of the electron-phonon interaction; effects of anisotropy, resulting from the crystal lattice structure of the material, were neglected.

Starting from the BCS-theory, but generalized to handle spatial variations of the pairing potential by a Green's function formalism, Gor'kov /2/ was able to deduce the Ginzburg-Landau (GL-) equations for the case, that the temperature was just below T_c and the magnetic field varied slowly in space over a coherence length ξ_0. This strict limitation to $(T_c-T) \ll T_c$ implies $\lambda(T) > \xi_0$ (λ denotes the penetration depth), so that the electrodynamics become local and the expansion of the free energy is valid. The effective wave function $\psi(r)$ turns out to be proportional to the local gap parameter $\Delta(r)$.

The intermediate and mixed states, in which the magnetic field enters in a non-perturbative way, typically involve a spatial inhomogeneity of the energy gap parameter Δ in the material. The structure of the mixed state (Shubnikov state) was calculated on the basis of the GL-equations. Abrikosov /3/ obtained an analytical solution for arbitrary κ ($\kappa = \lambda(T)/\xi(T)$) near H_{c2} and numerical solutions for large κ near H_{c1}. In thermodynamic equilibrium the stable configuration is found to be a triangular lattice /4,5/ of quantized flux-enclosing supercurrent vortices. Solutions of the Gor'kov equations for the ideal flux line lattice (FLL) in various special cases exceeding the GL-limit are summarized in /6/. They are nearly complete for isotropic superconductors in the clean ($\alpha \to 0$) and dirty ($\alpha \gg 1$) limit ($\alpha = 0.882\ \xi_0/\ell$ denotes the impurity parameter and ℓ the electron mean free path). From the Gor'kov equations /2/, which contain only an isotropic energy gap, no effects of anisotropy arise.

Neutron diffraction studies provided the first direct proof of the FLL in a type-II superconductor /7/. Today, neutron physical methods are able to give information on the morphology of a FLL, its orientation relative to the crystal lattice (CL), and the microscopic field distribution (for a review cf. /8/). The decoration technique of Träuble and Essmann /9,10/, on the other hand, first provided detailed information on individual flux lines and their positions, and experiments with this technique /11,12,13/ initiated all the work on the special features of low-κ superconductors.

In this paper we present some results on magnetic structures at the surface of low-κ Pb-Tl/In and Nb anisotropic materials at low temperatures revealed by the decoration technique.

2. Experiments on Low-κ Type-II Superconductors

2.1 Intermediate-Mixed State

Before flux penetration occurs a type-II superconductor distorts the external field H_e in the same way as a type-I superconductor. With the demagnetization factor D in field direction the sample will enter the mixed state when H_e is equal to $(1-D)H_{c1}$. In ideal specimens with $\kappa \gg 1$ an increase of magnetic field should produce a gradual growth of the mixed state without any need for an "intermediate state" in the form of a mixture of Meissner and Shubnikov phases. For small κ, however, an "intermediate-mixed-state" was observed within the field interval $(1-D)H_{c1} < H_e < (1-D)H_{c1} + D(B_0/\mu_0)$, loosely referred to as a "long range attractive interaction" between flux lines balancing the repulsive Landau interaction at an equilibrium distance a_{oo} /11/. Using Sharvin's approach Krägeloh /13/ determined the wall energy between the Meissner phase and the flux carrying domains in Pb-Tl polycrystals as a function of the GL-parameter κ and calculated the "spontaneous induction" B_0 from the observed flux line density B' at the specimen surface, using Landau's branching model. His results are shown in Fig.1. Obviously, the attractive interaction between flux lines becomes negligibly small

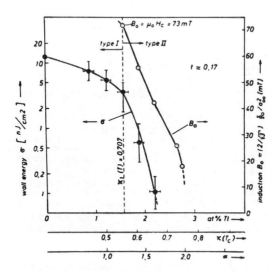

Fig.1: Induction B_0 and wall energy σ_{MF} between Meissner phase and flux-carrying regions in Pb-Tl alloys as function of the GL-parameter ($t \approx 0.17$), /13,6/

for impurity parameters $\alpha = 0.882\ \xi_0/\ell > 2$. Wall energies in superconducting Ta and Nb have been reported by Essmann /14/ recently.

The calculated $B_0 = (2\sqrt{3})(\phi_0/a_{oo}^2)$ is equivalent to the jump in the magnetization curves (for $D = 0$) at H_{c1}. This has been confirmed by Kumpf /15/ in the system Pb-Tl and by Auer and Ullmaier /16/ in Ta doped with nitrogen. Evidently, type-II superconductors with $\kappa(T_c) < 1.2$ show a first order phase transition at H_{c1}, at least at low temperatures.

2.2 Microscopic Anisotropies

A special feature of these materials with low Ginzburg-Landau and impurity parameter, respectively, lies in the fact that the response of the superconductor to an external magnetic field depends on the structure of the crystal lattice with respect to the field direction. The symmetry of the FLL fits the symmetry of the crystal axis parallel to the applied field, and there is a strong correlation between the orientations of the FLL and the anisotropic CL; an example is given in Fig.2. In the case that the field is applied

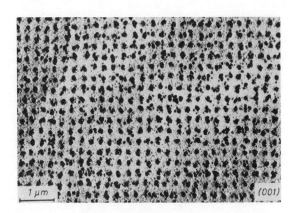

Fig.2: Flux line lattice in a Pb-Tl single crystal with the applied field in <001> direction; the nearest-neighbor directions are parallel to <100> crystal directions. ($\kappa \approx 0.72$, $t = 0.17$; $H_e: 0 \to H_{c2} \to 365$ Oe).

in the <100> direction, there is, under special circumstances, a lack of uniqueness with regard to the coupling of these two lattices. Fig.3b illustrates with the example of Nb, that two square lattices exist in the sample, which are tilted by an angle of 32° with respect to each other, the volume portion occupied by each lattice being about the same /20/. The nearest-neighbor directions of one square lattice are parallel to <110>, whereas the other sides can not be correlated to low-index crystal axes. Schelten et al. /19/ deduce from neutron diffraction experiments that at higher temperatures (T ≥ 4.2 K) and for higher flux densities ($B > B_o$) the morphology of the FLL changes to two distorted triangular lattices.

A similar phenomenon of two interpenetrating square lattices at low temperatures (T = 1.2 K), with the applied field in a <100> direction, has been observed in Pb-Tl alloys, too, although very seldom /21/. In contrast to Nb, however, the two lattices are tilted by an angle of 45° with respect to each other, the nearest-neighbor directions being parallel to <100> and <110>, respectively.

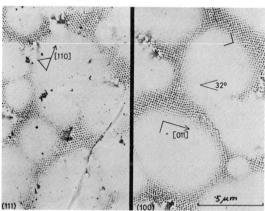

Fig.3: Intermediate-mixed state of (111) and (100) oriented Nb-discs, respectively. (t = 0.17, $(NI)_H$ transition with H_e = 600 Oe, D ≃ 1), resistivity ratio $R_{293}/R_{4.2}$ > 1000. (Courtesy of Dr.U. Essmann, Max-Planck-Institut für Metallforschung, Stuttgart.)

With increasing GL-parameter (i.e. with increasing Tl-content in Pb-Tl alloys) the influence of the CL-anisotropy decreases. There is a gradual transition from the "anisotropic FLL" to the triangular lattice as predicted by the GL-theory for an isotropic superconductor, the transition being complete at $\kappa > 1$.

The investigations on the morphology of flux line crystals have been extended by neutron diffraction studies to Tc and Pb-Bi /19/.

2.3 Macroscopic Anisotropies

Under equivalent conditions (κ, α, D, t, H) single crystalline discs establish an intermediate-mixed state structure in analogy to polycrystals. Contrary to that, however, diamagnetic regions are imbedded in the Shubnikov phase and arranged over macroscopic distances, thus forming a kind of "superlattice" (SL) (Fig.4). These struc-

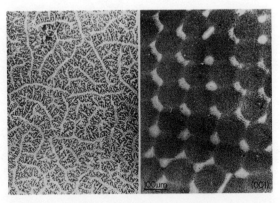

Fig.4: Intermediate-mixed state in Pb-Tl ($\kappa \approx 0.72$, t = 0.17, D $\approx$ 0.7, H_e: $0 \to H_{c2} \to 210$ Oe): a) polycrystal b) single crystal with the field parallel to <001>, the nearest-neighbor directions being parallel to <100> (dark: Shubnikov phase; optical microscope-photograph)

tures are different for different crystal orientations, and in the plane perpendicular to the field they adopt the symmetry of the crystal lattice in the direction of the magnetic field. The SL is (in most cases) correlated to the CL in a characteristic way. In Pb-Tl alloys the lattice-"points" itself are complicated agglomerations of small Shubnikov and Meissner domains (cf. Fig.5); there is a marked difference to the SL of Pb-In alloys /18/.

The existence of flux-free regions with a well-developed "macroscopic anisotropy" does not affect fundamentally the correlations of the FLL with the CL.

In analogy to the behavior of the FLL the "<100>-super-lattice" shows two preferred orientations relative to the crystal lattice (at least in Pb-Tl at low temperatures) with the same correlations as observed in the FLL. (The same observation was also made for $\kappa < 1/\sqrt{2}$). Both square lattices appear more or less extended with equal fre-

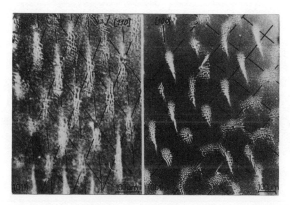

Fig.5: Superlattices in Pb-Tl single crystals ($\kappa \approx 0.72$, $t = 0.17$, $D \approx 0.5$) with long range interactions in <110>. (Left: (011)-disc, H_e: $0 \rightarrow H_{c2} \rightarrow 70$ Oe; right: (001)-disc, H_e: $0 \rightarrow H_{c2} \rightarrow 210$ Oe; dark: Shubnikov phase)

quency, and Fig.6 gives a statistical representative example. In addition, this micrograph illustrates (on the right hand side) that the square-SL has a certain "intrinsic stability": the symmetry of the CL in the direction of the external field induces the fourfold symmetry of the SL, but there is no longer any correlation between the two lattices. Observations of this kind have only been made on (001) planes perpendicular to the field; the orientations of the two- and threefold superlattices are strongly correlated to the CL.

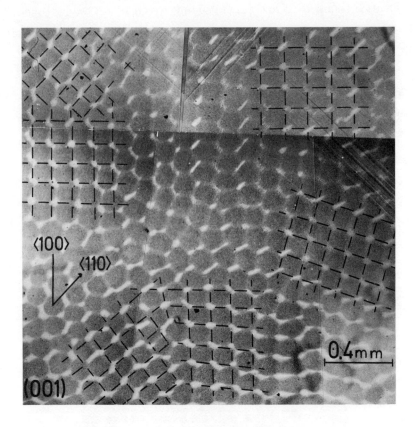

Fig.6: Superlattice in a Pb-Tl single crystal ($\kappa \approx 0.72$, $t \approx 0.17$, $D \approx 0.7$; H_e: $0 \rightarrow H_{c2} \rightarrow 210$ Oe) with its <001> axis parallel to the applied field H_e (to be compared with Fig.3; dark: Shubnikov phase; optical microscope)

Let B' be the flux density emerging at the surface of the Shubnikov domains. As Krägeloh /13/ has pointed out with the example of Pb-Tl, B' depends on the volume portion of the specimen being in the Shubnikov phase, i.e. $B' = B'(D, H_e)$; in the interior of the specimen the vortex lattice spacing a_{oo} is constant as long as there are flux-free regions at all (cf. section 2.1). For equivalent geometries, we always find $B'_{poly} > B'_{(001)}$ with a field dependence analogous to that of the polycrystalline material /18/. Under the condition that Landau branching in single crystals is equivalent to branching in polycrystals, this would imply $B^o_{poly} > B^o_{(001)}$.

What about the field dependence of the SL? In principle two possibilities exist for the intermediate-mixed state to respond to changes of the external field:
- to change the lattice spacing, i.e. the number of Meissner regions
- to change the dimension of the Meissner region and keep the lattice spacing constant.

According to our observations the spacing of the SL is essentially constant (of the order of 100 μm for $D \simeq 0.7$ and $\kappa \simeq 0.72$) in the entire field region $(1-D)H_{c1} < H_e < (1-D)H_{c1} + D(B_o/\mu_o)$ (cf. Fig.7), the influence of the field manifesting itself mainly in an extension of the Shubnikov phase. This result is the analogon to the flux line attraction with an equilibrium distance a_{oo} in the bulk. With increasing field H_e the substructure of the lattice-"points" - probably a consequence of anisotropic Landau branching at the specimen surface (see Fig.4) - shrinks to "stripes" or "spots" with only a few phase boundaries on a microscopic scale, the macroscopic arrangement transforming gradually into a threefold symmetry (Fig.7).

Starting from low fields again, i.e. from a square SL, we investigated the κ-dependence of the anisotropic intermediate-mixed state in Pb-Tl and Pb-In alloys. There is strong evidence that the lattice-"points" stretch to small Meissner domains (i.e. branching

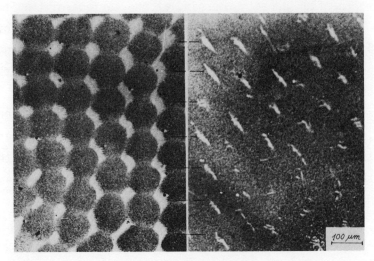

Fig.7: Superlattices in Pb-Tl ($\kappa \approx 0.72$, $t = 0.17$, $D \approx 0.7$, H_e parallel to <100> at different fields); left: $0 \rightarrow H_{c2} \rightarrow 210$ Oe with nearest-neighbor directions parallel to <100>, right: $0 \rightarrow H_{c2} \rightarrow 420$ Oe

disappears), the domains becoming smaller and smaller with increasing GL parameter, the SL disappearing in this way gradually. In spite of the transition in the FLL from the fourfold to the threefold symmetry (cf. section 2.2) the phase boundaries remain strongly correlated to crystallographic directions (cf. Fig.8).

Clearly, the presence of pinning centers tends to destroy the correlations between FLL and CL. With pinning, the equilibrium state in the specimen cannot be achieved, resulting in a lack of uniqueness in the B - H relation. Defects resist flux motion until the magnetic driving force P_D overcomes the pinning force P_V caused by the defects, and the condition $P_D = P_V$ defines a "critical flux gradient". Instabilities are a natural consequence of this "critical state": through thermal activation or external stimulation an abrupt motion of a large amount of flux may occasionally occur, resulting in a sudden loss of a large fraction of the specimen magnetization B. The power

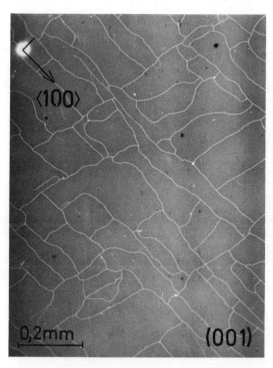

Fig.8: Phase boundaries following approximately <100> crystal directions with the flux line lattice in the Shubnikov phase (dark) having threefold symmetry.
(Pb-Tl, $\kappa \simeq 1$, $D \rightarrow 1$, $H_e : 0 \rightarrow H_{c2} \rightarrow 280$ Oe)

dissipation, inherent in this process, may lead through an increased local temperature to an increase of flux motion and still more power dissipation, and hence to a thermal instability; in Figs. 9 and 10 decorations of such instabilities are shown (Pb-Tl-(001), $\kappa \simeq 0.71$, T = 1.2 K, $D \simeq 0.6$, H_e = 350 Oe). Experiments of this kind demonstrate that the crystal anisotropy governs the flux motion on a macroscopic scale even in highly hysteretic superconductors, the flux flow occuring in the <110> directions on (001) oriented discs. <110> are the nearest-neighbor directions (i.e. the directions with long range interaction) in superlattices with two- and threefold symmetry, partially also when $\underline{H}_e$ is parallel to the fourfold crystal axis (cf. Fig.6). It is well known (cf. e.g. /9/) that in polycrystalline materials flux enters and leaves the specimen essentially in radial directions, i.e. perpendicular to the edge.

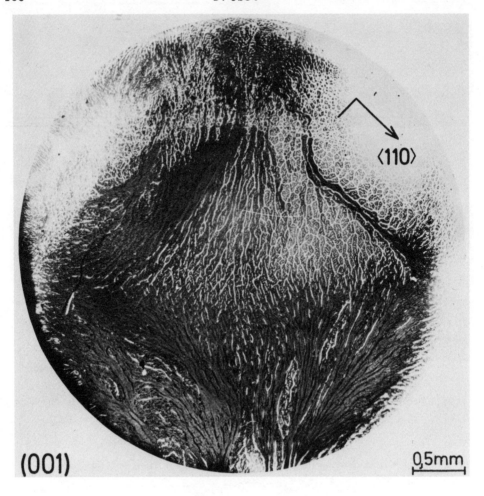

Fig.9: Flux motion, due to thermal instability, in a (001) oriented highly hysteretic disc, reflecting the fourfold crystal symmetry in direction of the applied field. (Pb-Tl, $\kappa \approx 0.71$, T = 1.2 K, $D \approx 0.6$; H_e: $H_{c2} \rightarrow 350$ Oe, bright: Meissner phase)

3. Summary and Discussion

Using pure Nb- and Pb-Tl/In-alloys as examples we have demonstrated, that two main groups of type-II superconductors have to be distin-

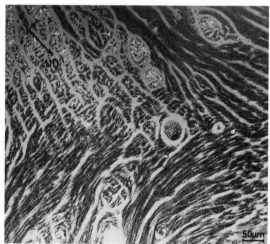

Fig.10: Flux motion in a hysteretic disc governed by the crystal anisotropy (to be compared with Fig.9). (Pb-Tl, $\kappa \approx 0.73$, T = 1.2 K, D $\approx$ 0.5; $H_e \parallel$ <100>: $H_{c2} \to 210$ Oe; specimen edge on the right hand side.)

guished with respect to their response to magnetic fields at low temperatures: low-κ materials ($\kappa \lesssim 1.2$), where the magnetic structures are determined by an effective flux line attraction and correlated to the crystal anisotropy in a characteristic way (type-II/1 superconductors), and impure superconductors with $\kappa > 1.2$, which can be described in good approximation by the GL theory (type-II/2 superconductors).

3.1 Flux Line Lattice Anisotropy

The symmetry of the FLL in type-II/1 superconductors fits the symmetry of the CL in the direction of the applied magnetic field H_e. For H_e parallel to <110> and <111> crystal directions, respectively, the orientations of the two lattices with respect to each other are correlated in a unique way. Evidently, there is a lack of uniqueness for H_e parallel to <100> (at least for finite D):

square lattices with two different orientations have been observed. In Pb-Tl the nearest-neighbor directions are predominantly parallel to <100>, and only very scarcely parallel to <110>; in Nb the two vortex lattices, tilted with respect to each other by an angle of about $32°$, exist with equal volume portions within the specimen (for $t \ll 1$), the nearest-neighbor directions of one of the two lattices being parallel to <110> (cf. Fig.3).

These distortions of the FLL from the triangular symmetry /3-5/ indicate an anisotropy of the "interaction potential" between vortices. Neither can they be understood in the framework of the GLAG theory nor with the generalization proposed by Ginzburg /22/ to substitute the isotropic electron mass by a tensor m_{ij} in the usual expression for the free energy. As Hohenberg and Werthamer /23/ pointed out, the effective mass is isotropic in cubic materials, and effects of anisotropy should reflect "nonlocal" features of the Gor'kov theory going beyond the scope of the GL theory. These nonlocal effects diminish with increasing temperature and impurity scattering, and the theory predicts a disappearance of the relative anisotropy.

There is experimental evidence /17,24/, however, that in addition to "nonlocality" other mechanisms may lead to a small anisotropy, which become effective in the nearly local limit. Numerous models have been proposed to describe, at least in a qualitative way, the FLL-anisotropy:
- a generalization of the London model with an anisotropic penetration depth λ, the principle source of the anisotropy being the phonon density of states /18,25/, and the electron density of states at the Fermi surface /26/, respectively;
- an extension of the Gor'kov equations in the limit $H \simeq H_{c2}$ and $T \simeq T_c$ with two anisotropy parameters to fit the orientation and structure of the FLL, the anisotropy originating from the Fermi

surface /27/.

None of these models deals with all aspects of the experimental observations.

Starting from Eliashberg's theory of the superconducting electron-phonon system Teichler recently /28,29/ gave a detailed treatment of cubic superconductors in an external magnetic field. According to this theory the microscopic anisotropies of the electron band structure, the phonon system, and the phonon-induced electron-electron coupling (resulting, in turn, in a wave-vector dependence of the energy gap) all contribute to macroscopic anisotropy phenomena. A unique description of the microscopic state with a single (complex) order parameter is not possible, as Teichler /28/ pointed out. The microscopic theory is valid in the limit of well-separated vortices (i.e. $B_0 \ll H_{c1}$). The weights, attached to the different anisotropy parameters, are temperature-dependent and different for anisotropies of H_{c2} and lattice correlations, respectively /30/. The elaborate theory of Teichler is able to predict the anisotropy of the FLL-orientations and brings about a lot of physical insight into these anisotropy phenomena.

An alternative explanation of the lattice correlations, neglecting all the microscopic anisotropy-parameters, was given by Ullmaier et al. /31/. According to this model certain orientations of the FLL with respect to an elastic anisotropic CL are favored due to the magneto-elastic strain field of the FLL. There is, however, little experimental evidence to confirm this idea; e.g. the transition from two square lattices in Nb (Fig.3) to two distorted hexagonal lattices /19/, when the temperature is raised from $t < 0.19$ to $t > 0.46$, can hardly be understood in terms of elastic interactions, since the elastic energy is negligible with respect to the magnetic energy as was shown by the authors of refs. /26/ and /32/.

3.2 Superlattice Anisotropy

In single crystals an intermediate-mixed state has been found which differs considerably from that in polycrystals (cf. Fig.4): isolated diamagnetic regions are imbedded in the Shubnikov phase in a regular manner, the order being of long range. We call this state a "superlattice" (SL).

In analogy with the FLL, the SL has the symmetry of the CL in the direction of the applied field H_e (Fig.5). For H_e parallel to <110> and <111>, respectively, the nearest-neighbor directions always are the <110> crystal directions; with H_e parallel to <100> two lattice orientations have been observed, <110> and <100> (compare Fig.4b with Fig.5b). In some manner, an "intrinsic anisotropy" is inherent to the square lattice as was shown in Fig.6. Do these observations permit the conclusion that the magnetic interaction is predominant?

The lattice-"points" of the SL at the surface of the specimen are agglomerates of small Meissner and Shubnikov domains (Fig.5). Following Krägeloh /13/ we interprete these complicated structures to be a consequence of Landau branching. This multiple branching is missing nearly completely in Pb-In alloys /18/, where the intermediate-mixed state extends to higher κ-values (>1) than in Pb-Tl. From both facts we deduce that the wall energy σ_{MF} in Pb-In is higher than in Pb-Tl (cf. Fig.1). Since in Pb-In-Tl-Bi alloys the main features of the pure-lead phonon spectrum are practically unchanged /33,34/, electronic contributions to σ_{MF}, and hence to the orientation of phase boundaries, must play an important role.

With an increase of field the multiple-branched lattice points shrink to small Meissner domains, the square lattice thereby transforming gradually to nearly threefold symmetry (Fig.7), whereas the FLL does not alter its symmetry. Changes in the portion of Shubnikov to Meissner phase primarily change the magnetic energy. This and the above mentioned fact that the square-SL is not necessarily correlated to CL-directions (and hence to the FLL) suggests to describe the anisotropic interaction potential with a London-like-model, originally proposed to describe the anisotropy of the FLL /18,26/.

The lack of correlation between the orientation of the FLL and the SL with respect to each other is not only expressed by the different field dependence of the two lattices but also by their different dependence on the GL parameter: the <001>-SL keeps its fourfold symmetry with increasing κ (Fig.8) indicating that the (microscopic) anisotropy parameters enter the anisotropies of the FLL and the SL, respectively, with different weights /28-30/. To understand the SL-anisotropy only in terms of the anisotropic magnetoelastic energy does not seem to be possible /35/.

3.3 Flux Motion Anisotropy

It has been demonstrated (Figs.9 and 10) that the CL anisotropy does not only affect the static arrangement of flux lines but also the flux line dynamics.

Flux motion is a dissipative process, the entropy of transport arising from vortex excitations. Does, therefore, any gap anisotropy Δ_k explain the experimental observations?* It is well known /36/, that in low-κ Pb-In alloys the thermal conductivity increases drastically, i.e. by a factor >2, in the temperature range from 1 to 3 K, and we expect a similar behavior for Pb-Tl. Does, perhaps, the anisotropic flux motion reflect an anisotropy of the energy release, i.e. the thermal conductivity? In Karlsruhe experiments are prepared to answer these questions.

---

*Note added in proof: With regard to this question cf. also papers R-6 (p. 213), C-13 (p. 257) and C-14 (p. 265).

The author wishes to thank Prof.Heinz for encouraging this work, Dr.U.Essmann for providing Fig.4, Dr.R.Meier-Hirmer for bringing ref./36/ to his attention, and Dr.G.Ries for helpful discussions on the subject of section 3.3.

References

/1/ J.Bardeen, L.N.Cooper, J.R.Schrieffer, Phys.Rev. 106, 162 (1957); 108, 1175 (1957)

/2/ L.P.Gor'kov, Sov.Phys.JETP 7, 505 (1958); 9, 1364 (1959); 10, 998 (1960)

/3/ A.A.Abrikosov, Sov.Phys.JETP 5, 1175 (1957)

/4/ W.H.Kleiner, L.M.Roth, S.H.Autler, Phys.Rev. 133, A1226 (1964)

/5/ J.Matricon, Phys.Lett. 9, 289 (1964)

/6/ U.Essmann, Proc. Int. Discussion Meeting on Flux Pinning in Superconductors, 23. - 27.9.1974, St.Andreasberg/Harz (Eds. P.Haasen, H.C.Freyhardt; published by Akademie der Wissenschaften Göttingen), p. 23

/7/ D.Cribier, B.Jacrot, L.M.Rao, B.Farnoux, Phys.Lett. 9, 106 (1964)

/8/ H.W.Weber, Atomkernenergie 25, 234 (1975)

/9/ H.Träuble, U.Essmann, phys.stat.sol. 18, 813 (1966)

/10/ H.Träuble, U.Essmann, Jahrbuch der Akademie der Wissenschaften, Göttingen, 1967

/11/ N.V.Sarma, Phil.Mag. 18, 171 (1968)

/12/ B.Obst, Phys.Lett. 28A, 662 (1969)

/13/ U.Krägeloh, phys.stat.sol. 42, 559 (1970)

/14/ U.Essmann, Proc. 14th Int.Conf.on Low Temperature Physics, Otaniemi, Finland, 1975 (Eds. M.Krusius, M.Vuorio, North Holland Publ.Comp., Amsterdam) 1975, vol.2, p. 171

/15/ U.Kumpf, phys.stat.sol.(b) 52, 653 (1972)

/16/ J.Auer, H.Ullmaier, Phys.Rev. B7, 136 (1973)

/17/ B.Obst, phys.stat.sol.(b) 45, 453 (1971)

/18/ B.Obst, phys.stat.sol.(b) 45, 467 (1971)

/19/ J.Schelten, G.Lippmann, H.Ullmaier, J.Low Temp.Phys. $\underline{14}$, 213 (1974)

/20/ U.Essmann, Phys.Lett. $\underline{41A}$, 477 (1972)

/21/ B.Obst, unpublished

/22/ V.L.Ginzburg, Zh.Eksp.i.Teor.Fiz. $\underline{23}$, 238 (1952)

/23/ P.C.Hohenberg, N.R.Werthamer, Phys.Rev. $\underline{153}$, 493 (1967)

/24/ N.Ohta, N.Tanaka, T.Ohtsuka, Phys.Lett. $\underline{49A}$, 363 (1974)

/25/ A.J.Bennett, Phys.Rev. $\underline{148}$, A1902 (1965)

/26/ M.Roger, R.Kahn, J.M.Delrieu, Phys.Lett. $\underline{50A}$, 291 (1974)

/27/ K.Takanaka, Progr.Theor.Phys. $\underline{46}$, 1301 (1971); $\underline{49}$, 64 (1973); $\underline{50}$, 365 (1973)

/28/ H.Teichler, Phys.Lett. $\underline{40A}$, 429 (1972); Phil.Mag. $\underline{30}$, 1209 (1974)

/29/ H.Teichler, Phil.Mag. $\underline{31}$, 775 (1975); $\underline{31}$, 789 (1975)

/30/ K.Fischer, Thesis, University of Stuttgart (1975)

/31/ H.Ullmaier, R.Zeller, P.H.Dederichs, Phys.Lett. $\underline{44A}$, 331 (1973)

/32/ E.Schneider, Diplomarbeit, University of Stuttgart (1974)

/33/ J.G.Adler, J.E.Jackson, B.S.Chandrasekhar, Phys.Rev.Lett. $\underline{16}$, 53 (1966)

/34/ R.C.Dynes, J.M.Rowell, Phys.Rev. $\underline{B11}$, 1884 (1975)

/35/ G.K.Kornfeld, K.Takanaka, Phys.Lett. $\underline{55A}$, 417 (1976)

/36/ A.K.Gupta, S.Wolf, Phys.Rev. $\underline{B6}$, 2595 (1972).

C-9   EXPERIMENTS ON THE CORRELATION BETWEEN THE FLUX LINE LATTICE AND THE CRYSTAL LATTICE IN SUPERCONDUCTING LEAD FILMS

W. Rodewald

Optisches Institut der Technischen Universität

D-1 Berlin 12, West Germany

The properties of pure superconductors with a cubic structure depend on the crystal orientation /1,2,3/. Consequently the arrangement of normal domains in the mixed state or in the intermediate state should be correlated with the crystal lattice. This was predicted theoretically /4/ and demonstrated experimentally /5 - 9/. In the following, the effect of the crystal symmetry on the arrangement of vortices and normal lamellae in thin pure Pb-crystals will be reported.

In a perpendicular magnetic field the magnetic flux penetrates films of type I superconductors, e.g. Pb, in the form of vortices /10,11,12/. Each vortex should carry the total flux of one quantum ϕ_0 and the vortices are expected to be arranged in the form of a lattice /13,14,15/. For increasing film thickness the vortices conglomerate to spots and finally form normal lamellae /13/. Hence, in thin Pb-crystals the correlation between the crystal lattice and the vortex lattice or normal lamellae could be investigated.

The magnetic flux in superconductors was observed with high resolution by the decoration technique developed by Essmann and

Träuble /16/. At low temperatures the magnetic strayfields of the normal domains were marked by small iron particles. After warming up, the samples were analyzed in a transmission or scanning-electron microscope.

By this decoration-microscopy the distribution of the vortices or lamellae /17/ was investigated with respect to crystal imperfections, like grain boundaries or other pinning centers. Simultaneously the orientation of large crystals of the sample could be determined by selected area electron diffraction. A comparison of the micrograph with the diffraction pattern permits to study the correlation between the crystal lattice and the vortex lattice /18/ or normal lamellae.

Thin Pb-crystals were prepared by recrystallization of evaporated Pb-films in the electron beam of a transmission electron microscope. All samples were cooled down to 1.2 K in a perpendicular magnetic field. Figure 1 shows the vortex distribution on a 150 nm thick Pb-crystal in a perpendicular field of 320 A/cm (40 mT). Since

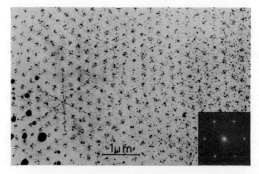

Fig.1: Triangular vortex lattice in a 150 nm thick Pb-crystal in a perpendicular field of 320 A/cm (40 mT) at $T = 1.2$ K. To improve the contrast of the vortices marked by small iron particles the Pb-film was burned away after having taken the selected area diffraction pattern (insert). A comparison of the micrograph with the diffraction pattern demonstrates the coincidence of orientation between the FLL and the CL.

the contrast of the small iron particles was very poor, the Pb-film was burned away in the beam of the electron microscope after having taken the selected area diffraction pattern of the crystal. The vortices are arranged in a triangular lattice with a period D = (240 ± 10) nm. The selected area diffraction pattern shows, that the crystal has a triangular symmetry too, and is oriented in the <111> direction. Taking the image rotation of the micrograph by the electromagnetic lenses into account, the orientations of the crystal lattice and of the vortex lattice coincide within the experimental error of about $10°$.

According to the theory of Lasher /15/ the densities of the free energy of a triangular and of a square vortex lattice do not differ very much. Hence, in crystals with a square symmetry a square vortex lattice is to be expected.

The preparation of thin crystals of square symmetry proved to be very difficult. Only in thick recrystallized Pb-films the crystal orientation sometimes differs from the <111> direction.

The vortex distribution in a 240 nm thick Pb-crystal having a <332> orientation is shown in Fig.2. The sample was decorated in a perpendicular magnetic field of 80 A/cm (10 mT). To improve the contrast of the vortex arrangement the Pb-film was burned away.

The vortices are aligned in vertical files, but perpendicular to these files no long-range order can be detected. Indexing the selected area diffraction pattern shows, that the vertical vortex files coincide with the (113) lattice planes, whereas the horizontal (220) lattice planes hardly cause any alignment of the vortices. This effect may be due to the different density of Pb-atoms in these crystal lattice planes and to the mixture of singly and doubly quantized vortices depending on the thickness of the crystals.

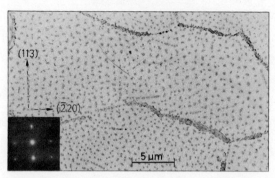

Fig.2: Vortex distribution in a 240 nm thick Pb-crystal in a perpendicular field of 80 A/cm (10 mT) at T = 1.2 K. The Pb-crystal was burned away after having taken the selected area diffraction pattern (insert). Indexing the diffraction pattern reveals the correlation of the vertical vortex-files with the (113) crystal lattice planes. There is hardly any correlation with the ($\bar{2}$20) crystal lattice planes.

These two experiments demonstrate that in nearly faultless superconducting crystals of cubic structure the vortices are correlated with the crystal lattice planes. In thicker Pb-crystals single vortices conglomerate to spots, as can be seen in Fig.2, and finally form lamellae.

The scanning electron micrograph (Fig.3) of a Pb-crystal of non-uniform thickness decorated in a perpendicular magnetic field of 240 A/cm (30 mT) shows the agglomeration of single vortices to lamellae. They are oriented in two preferred directions and cut each other under an angle of about 100°. The selected area diffraction pattern (insert of Fig.3), demonstrates, that this crystal is oriented in a <510> direction and hence is tilted about 11° from the <010> direction. The lamellae coincide with the (15$\bar{1}$) and (002) lattice planes of the crystal. The calculated angle between these lattice planes amounts to 101.1° and fits the angle between the lamellae quite well.

Figure 4 shows a scanning transmission micrograph of a 290 nm

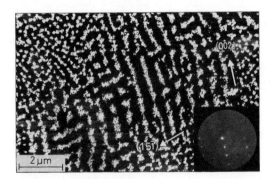

Fig.3: Scanning electron micrograph of Pb-crystals of non-uniform thickness decorated in a perpendicular field of 240 A/cm (30 mT) at T = 1.2 K. The vortices conglomerate to lamellae, which are correlated with the (151) and (002) lattice planes of the crystal. The insert gives the selected area diffraction pattern.

thick Pb-crystal decorated in a perpendicular field of 240 A/cm (30 mT). The normal lamellae can be detected as dark stripes on the crystal. They are arranged parallel to each other with a period of 650 nm. Often the lamellae are aligned along the grain boundaries. The selected area diffraction pattern (insert of Fig.4) indicates that the crystal is oriented in a <111> direction and that the

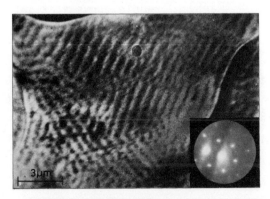

Fig.4: Scanning transmission electron micrograph of a 290 nm thick Pb-crystal decorated in a perpendicular field of 240 A/cm (30 mT) at T = 1.2 K. The normal lamellae (dark stripes) are arranged parallel to each other and are aligned with the crystal lattice planes of the <111> oriented Pb-crystal. The insert gives the selected area diffraction pattern.

lamellae coincide with the direction of the lattice planes. The branching angle between the lamellae amounts to about 120° and reflects the triangular symmetry of the crystal.

Summarizing the experimental results, they indicate that in nearly faultless Pb-crystals the arrangement of vortices, spots or lamellae is correlated with the crystal lattice planes.

I would like to thank Prof.H.Niedrig very much for his comments, Dipl.Phys.I.Richter (Inst.f.Metallforschung) for taking the scanning micrographs, and for a grant of the ERP fund.

References

/1/ P.C.Hohenberg, N.P.Werthamer, Phys.Rev. 153, 493 (1967)
/2/ K.Takanaka, Progr.Theor.Phys. 46, 357 (1971)
/3/ H.Teichler, Phil.Mag. 31, 775 (1975); 31, 789 (1975)
/4/ K.Takanaka, Progr.Theor.Phys. 49, 64 (1973); 50, 365 (1973)
/5/ B.Obst, phys.stat.sol.(b) 45, 453 (1971); 45, 467 (1971)
/6/ B.Lischke, W.Rodewald, Phys.Lett. 39A, 321 (1972)
/7/ A.Bodmer, U.Essmann, H.Träuble, phys.stat.sol.(a) 13, 471 (1972)
/8/ A.Bodmer, phys.stat.sol.(a) 19, 513 (1973)
/9/ H.W.Weber, J.Schelten, G.Lippmann, J.Low Temp.Phys. 16, 367 (1974)
/10/ M.Tinkham, Phys.Rev. 129, 2413 (1963)
/11/ H.Boersch, U.Kunze, B.Lischke, W.Rodewald, Phys.Lett. 44A, 273 (1973)
/12/ G.J.Dolan, J.Low Temp.Phys. 15, 111, 133 (1974)
/13/ A.L.Fetter, P.C.Hohenberg, Phys.Rev. 159, 330 (1967)
/14/ K.Maki, Ann.Phys. 34, 363 (1965)
/15/ G.Lasher, Phys.Rev. 154, 345 (1967)
/16/ H.Träuble, U.Essmann, phys.stat.sol. 18, 813 (1966)
/17/ B.Lischke, W.Rodewald, phys.stat.sol.(b) 63, 97 (1974)
/18/ W.Rodewald, Phys.Lett. 55A, 135 (1974).

C-10   ANISOTROPY IN THE INTERMEDIATE STATE
        OF SUPERCONDUCTING MERCURY*

R.P.Huebener[+], R.T.Kampwirth[+] and D.E.Farrell[x]

[+]Argonne National Laboratory, Argonne, Illinois 60439, USA

[x]Case Western Reserve University, Cleveland, Ohio, USA

1. Introduction

Anisotropy of the domain structure in the intermediate state of type-I superconductors has first been observed in single-crystalline disks of lead-thallium alloys of several mm thickness /1/. Subsequently, similar observations of anisotropy in the intermediate state structure were reported for single-crystalline foils of lead and tin with thicknesses between 6 and 100 μm /2,3/. A similar correlation between the orientation and structure of the flux line lattice has also been observed in type-II superconductors, and several papers are given on this subject at the present conference. The anisotropy in the magnetic domain structure is, of course, related to the corresponding anisotropy of one or more fundamental superconducting parameters, like the order parameter. Theories have been proposed to explain the anisotropy in the magnetic flux structure of the superconductor in terms of anisotropy of the Fermi surface /4/ and of the elastic properties /5/ of the crystal.

---

*Work supported by the U.S. Energy Research and Development Administration

In the present paper we report on recent magneto-optical observations of distinct anisotropy in the domain structure of the intermediate state in single-crystalline mercury. The sample thickness ranged from 50 to 1600 μm.

2. Mercury Samples

Our samples were prepared using triply distilled mercury with metallic impurities less than 5 ppm. A drop of mercury was placed on the aluminium-backed magneto-optic film used for flux detection through Faraday rotation of a beam of polarized light. The details of the magneto-optic system are described elsewhere /6/. A stainless steel plunger with a weight of 5 g was placed on top of the mercury drop. In this way sufficiently flat and thin, more or less circular mercury plates could be obtained. Following this arrangement of the sample, the apparatus was cooled through the melting point of Hg at the rate of about 1 °C/min. Through this simple procedure, single-crystalline plates of mercury can easily be obtained /7/. The single-crystalline nature of our samples could be deduced from the observed intermediate state patterns. However, no attempt was made to determine the crystallographic orientation of our samples (such an insitu x-ray study would have presented severe technical problems). We have investigated four crystals with thickness 47, 106, 360, and 1600 μm, all at the temperature $T = 1.7$ K ($t = 0.41$). A magnetic field could be applied perpendicular to the sample. The sample thickness was obtained for the thickest sample from the displacement of the plunger. The thickness of the 47 μm sample was found from the diameter of the original Hg drop (assumed spherical) and the final sample area. For the intermediate thicknesses a combination of weighing and area measurements was employed. Our thickness measurements are believed to be accurate within 10%. Mercury crystallizes in the lattice type A 10, i.e., in a rhombohedrically deformed fcc structure, the angle between the axes being approximately 70° /8/.

3. Magnetic Domain Structure

In Fig.1 we show as a typical case the domain structure in the Hg sample with 106 μm thickness at a magnetic field of 233 G applied perpendicular to the sample. This field value was set by monotonically decreasing H from a value just below H_c. Nearly the total circumference of this roughly elliptical sample is included in the photograph. The preferred domain orientation can clearly be seen. A more detailed view of the domain structure for the same sample is given in Fig.2. Distinct anisotropy of the domain orientation, such as seen in Figs.1 and 2, could be observed in all four specimens. Evidence for branching occured in the two thickest samples, with 360 and 1600 μm thickness. In the two thinnest samples branching was absent at all magnetic fields.

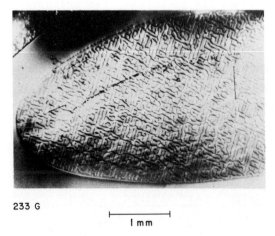

233 G

1 mm

Fig.1: Domain structure in a mercury plate with 106 μm thickness at 233 G. The field was reduced monotonically from just below H_c. Superconducting phase is dark.

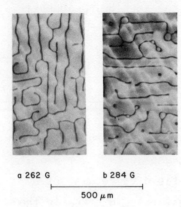

Fig.2:
Superconducting domains (dark) in the same sample as shown in Fig.1. (a) H = 262 G, decreased monotonically from just below H_c. (b) H = 284 G, increased monotonically from zero.

a 262 G   b 284 G
500 μm

The most regular domain patterns were obtained by first increasing the field to a reduced field value h ≈ 0.99 so that only a few isolated, nearly circular superconducting islands remained in the samples. By subsequently reducing the field the superconducting islands developed into straight long superconducting laminae, spanning often the entire length of the specimen. Upon reducing the field the growth of the superconducting phase always proceeded through the introduction of new domains from a variety of imperfections within the existing structure. Growth of the superconducting phase was never observed to proceed via fresh domains coming in from the edges and by crowding the existing domains closer together.

We note that the intermediate state pattern observed in our single crystal specimens and shown in Figs. 1 and 2 differs strongly from the completely arbitrary domain orientation usually observed in polycrystalline type-I superconductors in a perpendicular magnetic field. Highly anisotropic domain configurations in polycrystalline type-I specimens have only been observed previously in an inclined magnetic field with an appreciable field component parallel to the sample surface (Sharvin geometry) /9,10/.

Approaching the same field value by reducing the field from h ≈ 0.99 or by increasing the field from h = 0 resulted in somewhat different domain patterns. However, in both cases the anisotropy of the domain orientation was generally the same. In the field-decreasing case, for $0.25 < h < 1.0$ the domain pattern was found to display an open topology, i.e., no point in the sample was completely surrounded by superconducting phase. For these structures with open topology good reversibility was found as long as the field variation was restricted to the range $0.25 < h < 1.0$. These structures appear to approach the equilibrium configurations closely. Increasing the field from zero results in flux structures with closed topology, i.e., normal domains completely surrounded by superconducting phase. In this case, nearly circular normal flux spots often have been observed. A typical example is shown in Fig.3. We note that the flux spots are arranged in the form of highly regular but anisotropic arrays.

4. Conclusions

Our experiments on single-crystalline mercury plates in a perpendicular magnetic field clearly demonstrate a high degree of anisotropy in the domain structure of the intermediate state. This applies to the orientation of the rather straight superconducting and normal laminae as well as to the arrangement of flux-spot arrays.

103 G

200 μm

Fig.3: Intermediate state structure near the edge for the same sample as shown in Fig.1 at 103 G. The field was increased monotonically from zero. Normal phase is bright.

This anisotropic behavior is caused by the anisotropic interaction between the magnetic domains and the single-crystalline lattice. Mercury appears to be a convenient material for such observations, since single-crystalline specimens can easily be obtained by cooling the sample in-situ through the melting point of Hg. This ease of preparing single-crystalline specimens suggests furthermore to study the influence of controlled modulations at the sample surface on the structure of the intermediate state. (Such a modulation can simply be achieved by a proper treatment of the surface of the plunger used for flattening the Hg samples.) A correlation of the magnetic domain structure with in-situ x-ray studies appears to be difficult but not impossible and seems worth to be tried.

References

/1/ B.Obst, phys.stat.sol.(b) 45, 453 (1971)
/2/ A.Bodmer, U.Essmann, H.Träuble, phys.stat.sol.(a) 13, 471 (1972)
/3/ A.Bodmer, phys.stat.sol.(a) 19, 513 (1973)
/4/ K.Takanaka, Prog.Theor.Phys. 50, 365 (1973)
/5/ H.Ullmaier, R.Zeller, P.H.Dederichs, Phys.Lett. 44A, 331 (1973)
/6/ R.P.Huebener, R.T.Kampwirth, V.A.Rowe, Cryogenics 12, 100 (1972)
/7/ G.B.Brandt, J.A.Rayne, Phys.Rev. 148, 644 (1966)
/8/ M.C.Neuburger, Z.Anorg.Allg.Chem. 212, 40 (1933)
/9/ Yu.V.Sharvin, Soviet Phys. JETP 6, 1031 (1958)
/10/ D.E.Farrell, R.P.Huebener, R.T.Kampwirth, Solid State Comm. 11, 1647 (1972)

Present address of R.P.Huebener: Physikalisches Institut der Universität Tübingen.

C-11 MIXED STATE ANISOTROPY OF SUPERCONDUCTING VANADIUM

F.K.Mullen, R.J.Hembach, R.W.Genberg

Adelphi University

Garden City, N.Y. 11530, U.S.A.

Using a null deflection torque magnetometer, we have studied single crystal spheres of superconducting vanadium at 4.2 K and found that components of magnetization exist perpendicular to the field direction. Figure 1 represents a typical plot of torque vs. field magnitude H with the amplitude being a function of orientation. Torque in the Meissner region is a geometric effect caused by deviations from sphericity and is well understood on the basis of electromagnetic theory /1/. In the mixed state the orientation dependence of the torque exhibits the symmetry of the crystal structure. More precisely, Takanaka and Nagashima /2/ (TN) have shown that this torque is related to the anisotropy of the Fermi surface and have derived results for $H \simeq H_{c2}$ and $T \simeq T_c$. We find the orientation dependence of the torque to be in good agreement with the results of TN.

In order to extend the analysis throughout much of the mixed state, we have developed a phenomenological equation representing the torque in the mixed state as due to an interaction between the applied field H and magnetic poles localized about <111> directions on the surface of our spherical samples. For the contribution to the torque from a pole distribution about a single <111> axis, we write

$$\tau_i = VHF(H,T)\cos^3\gamma_i \sin\gamma_i \qquad (1)$$

where $F(H,T)$ reflects the density of poles about the <111> direction, V is the volume of the sphere and γ_i is the angle between the <111> direction and the magnetic field direction.

In our experiments H is applied in a horizontal direction and torque is measured about the vertical axis as a function of the magnitude of H and its angular orientation in the horizontal plane. If a <111> axis makes an angle β_i with its projection on the horizontal plane, it is a simple exercise in geometry to show that the vertical component of torque generated from Eq.(1) is

$$\tau_{vi} = VHF(H,T)\cos^4\beta_i \cos^3\alpha_i \sin\alpha_i \qquad (2)$$

where α_i is the angle between H and the projection of the <111> axis on the horizontal plane.

The contributions of the four <111> axes are summed for each crystal orientation investigated and compared with experimental measurements. The results are shown in Figs.2 and 3. All experimental data are for a value of $H = 0.87\, H_{c2}$ and $t = 0.8$. The calculated curves were scaled to give the best match for torque amplitude for curve (a) and this scaling was then maintained for the other curves. We feel that the orientation dependence and relative amplitudes of these curves are in good agreement with experimental results.

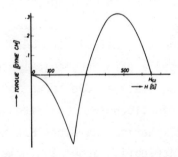

Fig.1: Typical torque versus magnetic field curve

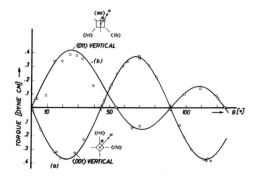

Fig.2: Torque versus field angle θ; (a) H in the (001) plane; (b) H in the (011) plane.

Curve (a) of Fig.2 represents torque versus field angle for the field H in the (001) plane and displays a $\sin 4\theta$ dependence, in agreement with the results based on Eq.(2). The direction of the torque is such as to cause the nearest horizontal <011> axis to align itself with the field. For curve (b) of Fig.2, the field lies in the (011) plane. The sense of the torque is reversed and is in such a direction as to cause rotation of the nearest horizontal <111> axis to align itself with the field direction. The experimental data show some scatter but the points at which the torque changes sign are well defined and in agreement with the results based on Eq.(2).

With the field in the (111) plane, the torque was observed to be small. Both the results of TN and those based on our phenomenological equation predict a null measurement. The nonzero torque may have arisen from a slight misorientation of the sample, from shape anisotropy or both.

For Fig.3 the sample was oriented such that the <111> axis was rotated approximately 7.5 degrees from the vertical toward an <001> direction. Based on Eq.(2), the predicted curve of torque versus field orientation is represented by curve (c), using the same scaling mentioned previously. Some scatter is observed in the experimental data, but the general agreement is clear.

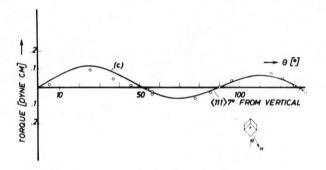

Fig.3: Torque versus field angle θ.

Using magnetization curves for the sample along with our torque data, we may also determine the effective angle ϕ between the net magnetization and the applied field. For this we use, of course, the expression

$$\tau = -4\pi \, V M \sin \phi \qquad (3)$$

Figure 4 represents a plot of ϕ versus H. At fields below H_{c1} ϕ is constant as is expected. Above H_{c1}, however, the results are somewhat surprising in that the curve is very nearly a linear function of the magnitude of the applied field throughout most, if not all, of the mixed state. It should be noted that the uncertainty in ϕ becomes large as H approaches H_{c2}, since both the torque and the magnetization approach zero. It is not clear from the data whether ϕ continues to increase linearly out to $H = H_{c2}$ or falls off as shown. The slope of the linear region exhibits the same orientation dependence as the torque curve; for example, corresponding to curve (a) $d\phi/dH$ exhibits a $\sin 4\theta$ dependence. The magnitude of the angle ϕ at H_{c2} can be obtained by extrapolation and for H in the (001) plane a maximum value of approximately 0.83 degrees is found for vanadium. For comparison purposes TN estimated a value of 0.87 degrees for niobium using anisotropy of H_{c2}. The closeness of these values might be expected since the Fermi surface of vanadium and niobium are similar.

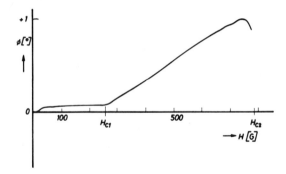

Fig.4: Angle φ between M and H versus magnetic field

As mentioned previously, the data near H_{c2} are consistent with the results of TN as well as those based on Eq.(2). Since the results of TN are limited to regions very near T_c and H_{c2}, the phenomenological model which assumes a surface pole density distributed about <111> axes, the net dipole moment of which has a $\cos^3$ dependence, is introduced to account for the torque data throughout much of the mixed state. For example, for H in the (001) plane, the orientation dependence of the torque varies very nearly as $\sin 4\theta$ for $H \geq 0.5$ H_{c2} which is explicable in terms of this model. This suggests to us that the flux line lattice is, in part, so oriented as to give rise to the pole distributions centered on <111> axes. On the contrary, if the flux line lattice is not so ordered but is directed in total along the field direction, then from symmetry no torque would exist. Since the anisotropy of the Fermi surface is the mechanism whereby the torque is generated, we suggest it may be the mechanism whereby the flux line lattice does not lie totally along the field direction.

References

/1/ J.A.Cape, J.M.Zimmerman, Phys.Rev. 153, 416 (1967)
/2/ K.Takanaka, T.Nagashima, Progr.Theor.Phys. 43, 18 (1970)

C-12 MEASUREMENT OF TORQUE DUE TO ANISOTROPY OF THE MAGNETIZATION VECTOR IN SUPERCONDUCTING NIOBIUM

R.Schneider, J.Schelten - Institut für Festkörperforschung der Kernforschungsanlage, D-517 Jülich, W-Germany and C.Heiden - Institut für Angewandte Physik der Universität, D-44 Münster, W-Germany

The anisotropy of the upper critical field in cubic anisotropic superconductors is well established. This is not the case for the anisotropy of the magnetization vector $\underline{M}$ in the mixed state. Although this phenomenon has already been predicted by Takanaka and Nagashima /1/ in 1970, the first successful experiment was reported only recently /2/.

The anisotropy of $\underline{M}$ involves that the vector $\underline{M}$ is in general not parallel to the internal field $\underline{H}$ and the component M_x perpendicular to $\underline{H}$ depends on the field direction. This is a consequence of the direction dependence of the free energy in anisotropic type-II superconductors. This direction dependence can be determined from the torque given by

$$D(\varphi) = -\frac{\partial E}{\partial \varphi} = \mu_0 \, V \, M \, H \sin \varepsilon_H \qquad (1)$$

where V denotes the sample volume and φ is defined in Fig.1. In order to determine the angle ε_H between the magnetization $\underline{M}$ and the internal field $\underline{H}$ torque experiments were made.

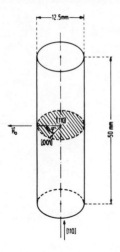

Fig.1: Sketch of the sample

The sample used was a monocrystalline niobium rod of 50 mm length and 12.5 mm diameter. A <110> symmetry direction was parallel to the cylinder axis. The sample was outgassed at 2300 °C for 26 h and a pressure of 2.6×10^{-10} torr and proved to be nearly reversible in a subsequent magnetization measurement. For the evaluation of ε_H, the magnetization curves of the outgassed sample and torque measurements were taken with the external field vector $\underline{H}_o$ in a (110) plane. All measurements were made at a reduced temperature of $t = 0.45$. The torque data were recorded with a torque magnetometer. Different torque curves were measured as a function of the applied field for various φ. The null deflection torque magnetometer which has been described by Fuhrmanns et al. /3/ provided a constant orientation of the sample with respect to the applied field during a specific experiment, i.e. φ was kept constant.

An example of the torque curves is given in Fig.2. It was obtained with the field $\underline{H}_o$ at an angle $\varphi = 43°$ off a <001> direction in the (110) plane. The figure shows that the torque in the mixed state is nonzero, decreases with increasing field and vanishes at the upper critical field H_{c2}. In the Meissner state the torque increases from 0 parabolically with increasing field. For an ideal

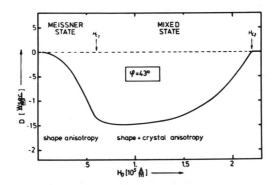

Fig.2: Torque D recorded by a null deflection torque magnetometer as a function of external field H_0 applied at an angle $\varphi = 43°$ (cf. Fig.1)

cylindrical sample the torque in the Meissner state would be 0. In our specimen the torque is caused by shape anisotropy. An analysis of the Meissner state data shows that they can be described satisfactorily by an ellipsoidal cross section with demagnetization coefficients $n_a = 0.494 \pm 0.002$ and $n_b = 0.506 \pm 0.002$ along two principal axes.

The torque in the mixed state consists of two contributions resulting from crystal anisotropy and shape anisotropy, respectively. Using the data of the Meissner state the shape anisotropy contribution in the mixed state was eliminated. It should be noted, however, that this contribution decreases rapidly with increasing field above H_{c1}.

After this correction for shape anisotropy the torque due to crystal anisotropy was recorded as a function of φ for fixed magnetic fields. As a typical result, the angular dependence at an applied field $H_0 = 1.43 \times 10^5$ A/m (179.4 mT) is shown in Fig.3. It can be seen that the torque vanishes if the applied field is parallel to any one of the symmetry directions <001>, <111> or <110>. A maximum of the torque occurs if the field is applied at an angle of 31° from a <100> crystallographic direction. Apart from these extrema side maxima

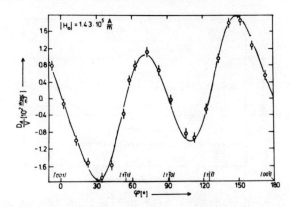

Fig.3: Angular dependence of the torque density D/V at H_o in the mixed state. The experimental data are fitted (solid line) by the function described in the text

appear. From the sign of the torque we conclude that the <111> and the <100> symmetry axes correspond to the easy and the hard direction of magnetization, respectively. The solid line in Fig.3 represents a least squares fit and was obtained by expanding the free energy of the mixed state by a four term series expansion using cubic harmonics $H_\nu(\varphi)$ with coefficients $K_\nu(H_o)$, which are functions of the external field only, i.e.

$$E(\varphi,H_o) = \sum_{\nu=0}^{3} K_\nu(H_o) H_\nu(\varphi) \qquad (2)$$

For the functions H_ν we use:

$$H_0 = 1 \qquad H_1 = \alpha^2\beta^2 + \alpha^2\gamma^2 + \beta^2\gamma^2$$
$$H_2 = \alpha^2\beta^2\gamma^2 \qquad H_3 = \alpha^4\beta^4 + \alpha^4\gamma^4 + \beta^4\gamma^4$$

where α, β and γ are the direction cosines of $\underline{H}_o$ with repect to the crystal axes. The fit function is then obtained from (2) using Eq.(1)

This series expansion fits the experimental data in a satisfactory way and yields the coefficients K_1 to K_3 as functions of the external field. The higher terms of this expansion contribute signi-

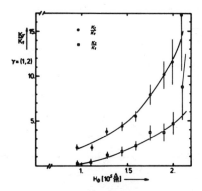

Fig.4: Field dependence of the coefficients K_2/K_1 and K_3/K_1

ficantly as is demonstrated in Fig.4, where the field dependence of the ratios K_2/K_1 and K_3/K_1 is shown. Over the entire field range K_2/K_1 is larger than 1 and increases up to a value of ≈ 15 near the upper critical field. K_3/K_1 is also a monotonically increasing function and larger than 1 except near H_{c1}.

Instead of discussing the absolute values of the anisotropy coefficients K_ν the angle ε_H, i.e. the angle between the magnetization and the internal field, will be considered. In Fig.5 the maximum of ε_H occuring at $\varphi = 31^0$ is shown as a function of H_0. This angle increases monotonically with increasing field and reaches an extrapolated value of 3.4^0 at H_{c2}. Theoretically /1/, ε_H near the upper critical field can be obtained from

$$\tan \varepsilon_H \approx 3H_{c2}(t) A(t) \frac{\langle v_x v_z v_\perp^2 \rangle}{\langle v_\perp^2 \rangle} \tag{3}$$

if the field is varied within a (100) plane. This expression contains components v_x, v_z and $v_\perp$ of the Fermi velocity, which are assumed to be temperature independent. Using Eq.(3) a theoretical value of 0.86^0 was predicted at a reduced temperature of $t = 0.7$. Taking into account the temperature dependence of $A(t)$ /1/ and of $H_{c2}(t)$ /4/ we calculate for $t = 0.45$ (where the torques were measured) a value of $\varepsilon_H = 3.3^0$ in excellent agreement with the measured angle.

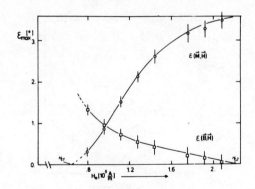

Fig.5: Field dependence of the maxima of the angles ε_H between $\underline{M}$ and $\underline{H}$ and ε_B between $\underline{M}$ and the induction $\underline{B}$

This experimental result is a strong indication that the anisotropy of the magnetization vector can be described by the theory of Takanaka and Nagashima.

We are grateful to Prof.W.Schmatz for his interest in this study and for valuable discussions.

References

/1/ K.Takanaka, T.Nagashima, Progr.Theor.Phys. <u>43</u>, 18 (1970)
/2/ R.J.Hembach, F.K.Mullen, R.W.Genberg, Proc. LT 14 (M.Krusius, M.Vuorio Eds.), North Holland 1975, Vol.2, p. 313
/3/ M.Fuhrmanns, C.Heiden, to be published
/4/ D.E.Farrell, B.S.Chandrasekhar, S.Huang, Phys.Rev. <u>176</u>, 562 (1968)

R-5 MICROSCOPIC CALCULATIONS OF ENERGY GAP ANISOTROPY[*]

J.P.Carbotte

Physics Department, McMaster University

Hamilton, L8S 4M1, Ontario, Canada

Abstract: In B.C.S. theory the possibility exists for anisotropy in the energy gap although there is no way to relate this anisotropy to the basic microscopic interactions that cause the superconductivity. To relate gap anisotropy to fundamentals we need to go to the Eliashberg gap equations. In this formulation the anisotropy is related to the
   a) Fermi surface
   b) the electronic wave functions
   c) the phonon spectrum
   d) the electron-phonon matrix element and in particular the Umklapp processes

all of which are anisotropic. We discuss these various sources of anisotropy and present results of theoretical calculations for the specific case of Al and of Pb.

1. Introduction

The possibility that the energy gap depends on the position of the electron on the Fermi surface already exists within the frame-

---

[*]Research supported by the National Research Council of Canada

work of B.C.S. theory. The gap $\Delta_{\underline{k}}$ for the state $\underline{k}$ is given by the standard B.C.S. equation /1/

$$\Delta_{\underline{k}} = \frac{1}{\Omega} \sum_{\underline{k}'} V_{\underline{k}',\underline{k}} \frac{\Delta_{\underline{k}'}}{2\sqrt{\varepsilon_{\underline{k}'}^2 + \Delta_{\underline{k}'}^2}} \qquad (1)$$

where Ω is the volume of the crystal and $\varepsilon_{\underline{k}}$ is the energy of the state $\underline{k}$ measured from the Fermi energy. The effective electron-electron potential represented by $V_{\underline{k}',\underline{k}}$ is assumed to be attractive and it is conventional to use a model in which it is taken to be zero unless both initial $\varepsilon(\underline{k})$ and final energies $\varepsilon(\underline{k}')$ are within $\pm \hbar\omega_D$ of the Fermi energy. Here $\hbar\omega_D$ is a typical phonon frequency. Further, if when both $\varepsilon_{\underline{k}}$ and $\varepsilon_{\underline{k}'}$ are within the rim shown in Fig.1, we take $V_{\underline{k}',\underline{k}}$ to be a constant V we get an isotropic gap for $|\varepsilon_{\underline{k}}| < \hbar\omega_D$ since (1) reduces to

$$\Delta_{\underline{k}} = V \frac{1}{\Omega} \sum_{\substack{\underline{k}' \\ |\varepsilon_{\underline{k}'}| < \hbar\omega_D}} \frac{\Delta_{\underline{k}'}}{2\sqrt{\varepsilon_{\underline{k}'}^2 + \Delta_{\underline{k}'}^2}} \quad \text{with } |\varepsilon_{\underline{k}}| < \hbar\omega_D \qquad (2)$$

and the R.H.S. does not refer to $\underline{k}$. On the other hand if we assume $V_{\underline{k}',\underline{k}}$ within the rim to be of the form /1/

$$(1 + a_{\underline{k}}) V (1 + a_{\underline{k}'}) \qquad (3)$$

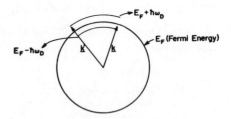

Fig.1: The B.C.S. model potential for electron scattering from initial state $\underline{k}$ to final state $\underline{k}'$. The width of the rim is of the order of a phonon energy.

where $a_{\underline{k}}$ is some anisotropy parameter we find instead that (1) gives

$$\Delta_{\underline{k}} = (1+a_{\underline{k}}) \{V \frac{1}{\Omega} \sum_{\substack{\underline{k}' \\ |\varepsilon_{\underline{k}'}|<\hbar\omega_D}} (1+a_{\underline{k}'}) \frac{\Delta_{\underline{k}'}}{2\sqrt{\varepsilon_{\underline{k}'}^2 + \Delta_{\underline{k}'}^2}}\}$$

(4)

with $|\varepsilon_{\underline{k}}| < \hbar\omega_D$

and hence $\Delta_{\underline{k}} = (1+a_{\underline{k}}) \times$ constant independent of $\underline{k}$ so that $\Delta_{\underline{k}}$ exhibits the same anisotropy as $V_{\underline{k},\underline{k}'}$.

It is important to realize that in this model no attempt is made to relate the effective potential $V_{\underline{k},\underline{k}'}$ to the fundamental parameters of the metal. Rather, the one parameter V which appears in the isotropic model is fitted to some experimental data, for example to the measured value of the gap Δ. In the anisotropic case not only must V be fitted, say to the average single crystal gap, but some model must also be assumed for the $\underline{k}$ dependence of the anisotropy parameter $a_{\underline{k}}$. Often if $a_{\underline{k}}$ is assumed small and the Fermi surface average of $a_{\underline{k}}$ taken to be zero it is not necessary to determine $a_{\underline{k}}$ in detail but one can get away with fitting a single parameter - the mean square anisotropy a^2 equal to the Fermi surface average of $a_{\underline{k}}^2$. If a^2 is obtained from some data thought to be dependent on the anisotropy one can then proceed to calculate other properties. This is discussed by the MIT group /2/.

In this talk I intend to describe how it is possible to go beyond BCS theory which of necessity deals with model interactions and treat accurately and in detail the real interactions that are responsible for superconductivity. This requires mathematics that is much more complex than that involved in BCS theory. Equation (1) gets replaced by the Eliashberg gap equations /3/. Before I describe these equations let me tell you where they come from and what they can do for us. They are fundamental to any theory which seriously attempts to relate the gap to the basic interactions in the metal.

To derive the Eliashberg gap equations one starts with the complete Hamiltonian for the coupled system of electrons and phonons. It consists of a single particle electron part containing the effects of the crystal potential, a phonon part dealing with the completely renormalized phonon frequencies, the electron-phonon interaction and the coulomb repulsion between the electrons. Thermodynamic Green's functions sufficiently general to include a superconducting solution are defined and a perturbation series generated for them. In the pure phonon model (in which case the coulomb repulsions are ignored) it is possible to sum the entire set of diagrams in the perturbation series that are thought to be numerically significant. This happy circumstance is a result of Midgal's theorem /4/ which states that corrections to the bare electron-phonon vertex are of order

$$\sqrt{\frac{m}{M}}$$

the square root of the electron to ion mass ratio. This implies that only the nested diagrams need be included to get numerically significant results. They can conveniently be summed by making use of the Dyson equation and the Nambu /5/ self consistency condition. More precisely, but still schematically, if we introduce a 2x2 matrix Green's function G with diagonal elements dealing with the ordinary Green's functions for spin up and down electrons and the off diagonal components dealing with the anomalous Gorkov amplitudes the Dyson equation is

$$G^{-1} = G_0^{-1} - \Sigma \tag{5}$$

where G_0 is the noninteracting electron Green's function and Σ is the electron self energy due to the interactions. Using a diagramatic notation the Nambu self consistency condition is simply

$$\Sigma = \begin{matrix}\text{(diagram)}\end{matrix} \qquad (6)$$

where the double solid line ═══▶═══ represents the fully interacting G, the solid dots the electron-phonon vertex and the wiggly line ∿∿ a phonon propagator. Iteration of (5) and (6) leads to the sum of nested diagrams for Σ, namely

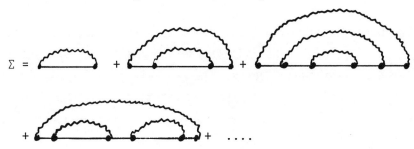

where the solid line ───▶─── represents the non interacting Green's function G_0. Equations (5) and (6) lead directly to the Eliashberg gap equations once the coulomb repulsions are included approximately.

2. Isotropic Eliashberg Equations

For an isotropic system at zero temperature the Eliashberg equations can be reduced to a one variable set of two non linear coupled integral equations for a complex frequency (ω) dependent gap $\Delta(\omega)$ and renormalization function $Z(\omega)$. They are /6/

$$\Delta(\omega)Z(\omega) = \int_0^{\omega_c} d\omega' \, \text{Re}\left\{\frac{\Delta(\omega')}{\sqrt{\omega'^2 - \Delta^2(\omega')}}\right\} [K_+(\omega,\omega') - \mu^*] \qquad (7)$$

and

$$\{1 - Z(\omega)\}\omega = \int_0^{\omega_c} d\omega' \, \text{Re}\left\{\frac{\omega'}{\sqrt{\omega'^2 - \Delta^2(\omega')}}\right\} K_-(\omega,\omega') \qquad (8)$$

with

$$K_{\pm}(\omega,\omega') = \int_0^\infty d\nu \; \alpha^2(\nu)F(\nu) \{ \frac{1}{\omega'+\omega+\nu+i0^+} \pm \frac{1}{\omega'-\omega+\nu-i0^+} \} \quad (9)$$

In (7) and (8) ω_c is a phonon cut off usually taken to be equal to 10 times the maximum phonon frequency in the metal which we denote by ω_0. It is important to note that the form of these equations is universal and that any reference to a particular material comes only through the two kernels $\alpha^2(\nu)F(\nu)$ and μ^*. The parameter μ^* deals approximately with the coulomb repulsions between the electrons and $\alpha^2(\nu)F(\nu)$ contains all the information on the phonon mediated part of the effective electron-electron interaction. Before specifying these kernels further I want to discuss superconducting tunneling experiments because they can be used to measure $\alpha^2(\nu)F(\nu)$ and μ^* quite directly. Also they provide in the case of Pb junctions strong evidence for the validity of the isotropic Eliashberg equations (7) and (8).

For simplicity consider tunneling between a normal metal and a superconductor. The basic idea is illustrated in Fig.2. Consider a superconducting film, the surface of which has been oxidized to give an oxide barrier of about 20 Å thickness. A normal metal film is then deposited on the other side of the oxide. The oxide layer is to be thick enough that electrons cannot diffuse from one side to the other. Quantum mechanically they can still tunnel. Referring again to Fig.2 we see a potential barrier representing the oxide

Fig.2: Schematic representation of a tunnel junction

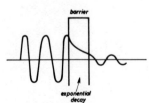

layer. On one side of the barrier is shown the wave function of an electron which decays exponentially in the barrier but still leaks out to the other side. There is a finite probability for an electron to tunnel from one side of the oxide layer to the other.

The current (I) flowing through the junction as a function of the voltage drop (V) across it contains a sharp and detailed image of $\alpha^2(\omega)F(\omega)$ and μ^*. This remarkable fact allows us to measure these two parameters and so obtain very valuable microscopic information on particular metals. The exact relationship between I as a function of V and $\alpha^2(\omega)F(\omega)$ is determined by the theory of superconductivity, more specifically through the Eliashberg gap equations. As we have seen, the Eliashberg gap equations are a complicated set of two coupled non-linear integral equations for a complex gap function $\Delta(\omega)$ and a renormalization function $Z(\omega)$. For the present purpose it is sufficient to know that Δ and Z can be used to compute many of the important properties of the superconducting state. For instance we can calculate from them the I-V characteristics of a tunneling junction.

For a given model of $\alpha^2(\omega)F(\omega)$ and an assumed value of μ^* we can solve the set of equations (7) and (8) and predict the I-V characteristic of the tunnel junction. This can be matched to the experimental results on I as a function of V and the kernel $\alpha^2(\nu)F(\nu)$ and μ^* adjusted until good agreement is obtained. This procedure is referred to as inversion of the Eliashberg equations. In Fig.3 we show a plot of $\alpha^2(\nu)F(\nu)$ obtained in this way for Pb by McMillan and Rowell /7/. The shape of $\alpha^2(\nu)F(\nu)$ is very similar to the shape of the phonon frequency distribution for Pb. Shortly I will describe how $\alpha^2(\nu)F(\nu)$ derived from tunneling compares with detailed first principle calculations of the same function. For the moment I wish instead to tell you how the tunneling data can be used to get an experimental verification that the Eliashberg equations are really quite accurate as claimed theoretically.

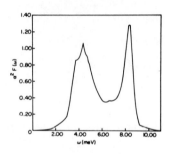

Fig.3: The isotropic spectral density $\alpha^2(\omega)F(\omega)$ for Pb obtained from inversion of the Eliashberg gap equations

The normalized conductance $\sigma(V)$ of a tunnel junction is defined as dI/dV in the superconducting state divided by its value in the normal state. Experimental results on $\sigma(V)$ vs V for a Pb-Pb junction are shown in Fig.4. As previously stated, the structure exhibited in this figure is an image of the structure in $\alpha^2(\omega)F(\omega)$. To invert the Eliashberg gap equations $\sigma(V)$ is needed only for values of $V \leq 11$ meV. This fact allows us to check experimentally the accuracy of the Eliashberg gap equations since once $\alpha^2(\omega)F(\omega)$ and μ^* are known we can solve the Eliashberg equations and predict $\sigma(V)$ beyond 11 meV. This prediction can be tested against experimental data /7/ as shown in Fig.5. (Strictly speaking it is not $\sigma(V)$ that is plotted but a closely related quantity.) The tremendous amount of agreement

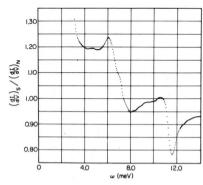

Fig.4: The conductance of a tunnel diode (involving Pb) versus voltage

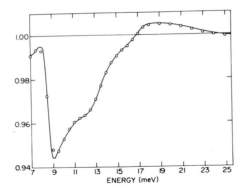

Fig.5: Comparison of theoretical and experimental I-V characteristic beyond 10 meV

found between theory (solid black curve) and experiment (open dots) is truly remarkable and leaves little doubt that equations (7) and (8) are accurate and therefore that tunneling data can be used to get accurate information on $\alpha^2(\omega)F(\omega)$ and μ^*. We are dealing here with a set of equations that represent an essentially exact solution to a many body problem.

More important in the present context is that these equations can be safely used to relate accurately superconducting properties to the basic microscopic interactions in the metal. A suitable generalized form of these equations to the anisotropic case forms a sound basis on which to begin a discussion of the dependence of gap anisotropy on the microscopic parameters.

We turn now to a discussion of the spectral density $\alpha^2(\nu)F(\nu)$ and of its dependence on the basic interactions between electrons.

3. Spectral Density $\alpha^2(\nu)F(\nu)$

The isotropic spectral density $\alpha^2(\nu)F(\nu)$ describes the phonon mediated part of the effective electron-electron interaction. It has incorporated in it all of the complicated information about the metal as it refers to superconductivity. That is it contains the information on the lattice dynamics, the electronic structure and the electron-phonon coupling. It can be visualized as is illustrated in Fig.6 in terms of two electrons scattering through the exchange of a phonon. The first electron polarizes the system of ions with the lattice polarization charge scattering the second electron. $F(\omega)$ can be thought of as the phonon frequency distribution and contains the information on the kinds of particles that can be exchanged while $\alpha(\omega)$ gives the strength at the electron-phonon vertex.

The formula for $\alpha^2(\nu)F(\nu)$ which contains no approximations /8/ is

$$\alpha^2(\nu)F(\nu) = \frac{\int \frac{dS_{k'}}{\hbar|V_{k'}|} \int \frac{dS_k}{\hbar|V_k|} \frac{1}{(2\pi)^3} \sum_\lambda |g_{k'k\lambda}|^2 \delta(\nu-\omega_\lambda(\underline{k}'-\underline{k}))}{\int \frac{dS_k}{\hbar|V_k|}} \quad (10)$$

with $\hbar$ Planck's constant divided by 2π, $\underline{V}_k$ the Fermi velocity for the state $\underline{k}$ on the Fermi surface and dS_k an element of Fermi surface area. The sum on λ extends over the various phonon branches and $\omega_\lambda(\underline{k})$ is a phonon frequency. Finally $g_{k',k\lambda}$ is the electron-phonon coupling which describes an electron scattering from $\underline{k}$ on the Fermi surface to $\underline{k}'$ on the Fermi surface due to the creation or absorption of a phonon of frequency $\omega_\lambda(\underline{k}'-\underline{k})$. It is important to stress that

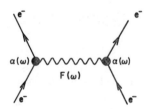

Fig.6: Schematic representation of the isotropic spectral density $\alpha^2(\omega)F(\omega)$

Microscopic Gap Calculations

(10) applies even when the electronic or phonon structure is very complicated. It is only through this function and through μ^* that the complexities associated with a particular metal enters the theory of superconductivity.

It is useful to contrast this function with the phonon frequency distribution $F(\omega)$ which can be written as

$$F(\omega) = \frac{1}{N} \sum_{\substack{k\lambda \\ F.B.Z.}} \delta(\omega-\omega_\lambda(\underline{k})) \qquad (11)$$

where N is the total number of ions in the system and $\underline{k}$ extends over the first Brillouin zone only (F.B.Z.) since the phonon frequencies are confined to this region. The only difference between (10) and (11) is that in (10) the phase space for the phonon lable $\underline{k}'-\underline{k}$ is obtained by all possible electron scattering from any initial state $\underline{k}$ on the Fermi surface to all possible final states $\underline{k}'$. Further, each phonon mode is now weighted by the square of the electron phonon coupling for that particular scattering instead of being given weight 1 as in $F(\omega)$.

For simple metals where the idea of a weak electron-ion pseudopotential W is valid, the electronic pseudo wavefunctions $\psi_{\underline{k}}$ can be sensibly expressed as a mixture of a few plane waves. We can write

$$\psi_{\underline{k}} = \frac{1}{\sqrt{\Omega}} e^{i\underline{k}\cdot\underline{x}} (\sum_{\underline{k}_n} e^{i\underline{k}_n\cdot\underline{x}} a_{\underline{k}_n}(\underline{k})) \qquad (12)$$

Here the $\underline{k}_n$'s are the inverse lattice vectors and the $a_{\underline{k}_n}(\underline{k})$'s the plane wave mixing coefficients. The sum over $\underline{k}_n$ can be limited to a few terms. For example in the case of Al, Ashcroft /9/ was able to fit the existing de Haas-van Alphen data on the geometry of the Fermi surface with a four plane wave pseudopotential model. In Fig.7 we show our results /10/ for the (001) cross section of the Al Fermi

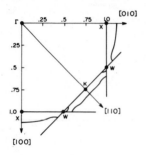

Fig.7:
The (001) cross section of the aluminium Fermi surface

surface following the work of Ashcroft. We note small but certainly significant distortions from a sphere. These deviations from sphericity will be shown later on to provide a source of gap anisotropy. For the moment however we are only concerned with the isotropic limit.

The electron-phonon coupling $g_{\underline{k},\underline{k}'\lambda}$ is given by the formula

$$g_{\underline{k}'\underline{k}\lambda} = -i\sqrt{\frac{\hbar\Omega}{2MN\omega_\lambda(\underline{k}'-\underline{k})}}\;\varepsilon_\lambda(\underline{k}'-\underline{k}).\{$$

$$\sum_{\underline{k}_n}\sum_{\underline{k}_{n'}} a^*_{\underline{k}_{n'}}(\underline{k}') a_{\underline{k}_n}(\underline{k})\,(\underline{k}'+\underline{k}_{n'}-\underline{k}-\underline{k}_n)<\underline{k}'+\underline{k}_{n'}|W|\underline{k}+\underline{k}_n>\}$$

(13)

where $\varepsilon_\lambda(\underline{k})$ is the phonon polarization vector associated with the $(\underline{k}\lambda)$'th mode. The factor $<\underline{p}'|W|\underline{p}>$ is the electron-ion pseudopotential form factor between two plane waves $|\underline{p}>$ and $|\underline{p}'>$.

While we have already referred to the case of Al it is convenient at this point to discuss Pb because tunneling results are available for $\alpha^2(\nu)F(\nu)$ in this case. To evaluate (10) numerically we have used the four plane wave model of Anderson and Gold /11/ for the Pb Fermi surface. This gives the Fermi surface elements $dS_{\underline{k}}$, the Fermi velocities $\underline{V}_{\underline{k}}$ and the plane wave mixing coefficients $a_{\underline{k}_n}(\underline{k})$. In addition we need the pseudopotential form factor $<\underline{p}'|W|\underline{p}>$ which enters explicitly expression (13). We take it from the work of Appapillia and Williams /12/. Finally, the lattice dynamics is needed. We take it from the Born von Karman force constant model of Cowley /13/. From the force constants the frequencies $\omega_\lambda(\underline{k})$ and polarization vectors $\underline{\varepsilon}_\lambda(\underline{k})$ follow quite directly at any point $\underline{k}$ in the F.B.Z.

In Fig.8 we compare our results for $\alpha^2(\nu)F(\nu)$ /14/ (solid line) with those of McMillan and Rowell /7/ (dashed curve) obtained from tunneling. The overall agreement is taken to be very satisfactory. The remaining small disagreements as to detail can probably be reduced by putting in slightly different input parameters in our calculations. This would be of no great significance at the moment.

We can take the good agreement found in Fig.8 as further evidence that the isotropic gap equations (7) and (8) are accurate. Further, and perhaps as important, it leaves little doubt that we are now able to calculate the kernels in these equations from fundamentals with considerable accuracy at least for the nearly free electron metals. This gives us confidence that we can make realistic calculations of gap anisotropy. To do so we will need to go to a generalized four dimensional form of the Eliashberg gap equations which specifically include a momentum label. Before doing this however we turn to a discussion of the kernels that will enter into these equations.

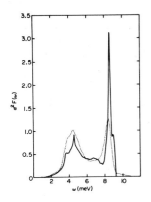

Fig.8:
Comparison of the results of first principle calculations (solid curve) for the isotropic spectral density $\alpha^2(\omega)F(\omega)$ with tunneling results (dotted curve)

4. Directional Spectral Densities $\alpha_{\underline{k}}^2(\omega)F_{\underline{k}}(\omega)$

To calculate the anisotropic energy gap at various points on the Fermi surface we will need to know, in lowest order approximation, the directional spectral densities /10/

$$\alpha_{\underline{k}}^2(\omega)F_{\underline{k}}(\omega) = \frac{1}{(2\pi)^3} \int \frac{dS_{\underline{k}'}}{\hbar |\underline{V}_{\underline{k}'}|} \sum_\lambda |g_{\underline{k}',\underline{k}\lambda}|^2 \delta(\omega-\omega_\lambda(\underline{k}'-\underline{k})) \qquad (14)$$

which refer to a particular electronic state $\underline{k}$ on the Fermi surface. The Fermi surface average of the $\alpha_{\underline{k}}^2(\omega)F_{\underline{k}}(\omega)$ gives the isotropic function $\alpha^2(\omega)F(\omega)$ which refers to all electrons participating in superconductivity i.e.

$$\alpha^2(\omega)F(\omega) = \frac{\int \frac{dS_{\underline{k}}}{\hbar |\underline{V}_{\underline{k}}|} \alpha_{\underline{k}}^2(\omega)F_{\underline{k}}(\omega)}{\int \frac{dS_{\underline{k}}}{\hbar |\underline{V}_{\underline{k}}|}} \equiv <\alpha_{\underline{k}}^2(\omega)F_{\underline{k}}(\omega)> \qquad (15)$$

where the brackets < > will from here on indicate a Fermi surface average. To get a feeling for the meaning of the directional spectral densities (14) we begin with a discussion of the renormalization of the electron mass due to the electron-phonon interaction. Figure 9 is a schematic representation of an electron propagating through a crystal lattice composed of ions of valence +Z. It polarizes the ions around it with the effect that a cloud of phonons

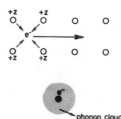

Fig.9: Schematic representation of the renormalization of the electron mass due to the electron-phonon interaction.

follows it in its motion. An exact formula for the ratio of the renormalized electron mass $m_{\underline{k}}^*$ to the bare mass is

$$\frac{m_{\underline{k}}^*}{m} = 1 + \lambda_{\underline{k}} \qquad (16)$$

where $\underline{k}$ is the electron label and $\lambda_{\underline{k}}$ the electron-phonon mass renormalization given by

$$\lambda_{\underline{k}} = 2 \int_0^\infty \frac{\alpha_{\underline{k}}^2(\omega) F_{\underline{k}}(\omega) d\omega}{\omega} \qquad (17)$$

i.e. it is simply equal to twice the first inverse moment of the directional spectral weight $\alpha_{\underline{k}}^2(\omega) F_{\underline{k}}(\omega)$ for the particular state $\underline{k}$.

The anisotropy in $\lambda_{\underline{k}}$ clearly reflects the anisotropy in these spectral weights.

The anisotropy in the distributions $\alpha_{\underline{k}}^2(\omega) F_{\underline{k}}(\omega)$ can be usefully discussed in terms of anisotropy in

1) the Fermi surface
2) the electronic wave functions
3) the phonon spectrum
4) the form of the electron-phonon interaction.

I have already shown a cross-section of the Fermi surface of Al and noted important distortions from a sphere. These distortions clearly lead to anisotropies. Also the electronic wave functions can differ very much from a plane wave $|k\rangle$ in regions where the Fermi surface is not spherical. Since this wave function enters as the initial state in the electron-phonon matrix element (13) differences will arise in $\alpha_{\underline{k}}^2(\omega) F_{\underline{k}}(\omega)$ from this source. Further, since the phonon frequencies $\omega_\lambda(\overline{k})$ will in general depend on the direction of $\underline{k}$ in the first Brillouin zone the phonon frequencies will also be a source of anisotropy in $\alpha_{\underline{k}}^2(\omega) F_{\underline{k}}(\omega)$. Finally we have anisotropy due to Umklapp processes. It is worth explaining this in some detail. To do so let us assume the Fermi surface to be a sphere and take the electronic wave functions to be plane waves. This is the free electron model in which band structure effects are totally ignored.

To illustrate the main features of the anisotropy that still remain, it is sufficient to consider a two-dimensional case with a square Brillouin zone (B.Z.) and a spherical F.S. that occupies half the B.Z. volume. In Fig.10(a) we show an electron at the Fermi surface in a definite initial state $|\underline{k}>$ which scatters to $|\underline{k}'>$. The energy transfer in the process is a phonon energy which is very much smaller than the Fermi energy and hence the phase space for scattering is limited to final states on the F.S. For a definite electronic transition from $\underline{k}$ to $\underline{k}'$, the momentum transfer $\underline{k} - \underline{k}'$ serves to label the phonon mode involved. For a given $\underline{k}$, the momentum $\underline{k} - \underline{k}'$ ranges over the sphere shown in Fig.10(b). It goes through the origin of the F.B.Z. and overlaps into higher B.Z. The parts in the higher zones give rise to Umklapp processes. To identify the phonons responsible in this case, it is necessary to map $\underline{k} - \underline{k}'$ back into the F.B.Z. This is illustrated in Fig.10(c). The solid curve in the F.B.Z. identifies the set of phonons that can scatter the electron in the state $|\underline{k}>$. Of course, if we choose a different initial state $|\underline{k}>$ we expect different phonons to be involved. This is shown in Fig.10(d) and 10(e). Not only is the set of phonons different but also the reciprocal lattice vectors involved in the Umklapp processes are quite distinct. These differences lead to large anisotropies.

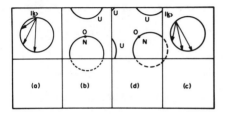

Fig.10:
Illustration of phonon and Umklapp anisotropy in $\alpha^2(\omega)F(\omega)$

Microscopic Gap Calculations

To present results we will need to specify some coordinate system with reference to which we can identify various directions of $\underline{k}$. Because of cubic symmetry the Fermi surface has 1/48'th symmetry and it is only necessary to have results in one such region. In Fig.11 we show a section of a sphere in the positive octant of a cartesian coordinate system x along <100>, y along <010> and z along <001>. The segment is further cut up into six regions by symmetry. In what follows we will present results for points in the region labeled III. To identify points on the Fermi surface we will use two polar angles θ and ϕ with θ measured from the z axis and ϕ in the x-y plane measured from the x axis.

In Fig.12 we show $\alpha_{\underline{k}}^2(\omega)F_{\underline{k}}(\omega)$ as a function of phonon frequency ω for $\underline{k}$ in three high symmetry directions, namely <100>, <110> and <111> /15/. Since these curves have been obtained assuming a free electron model the considerable amount of anisotropy exhibited is due only to Umklapp anisotropy and the anisotropy in the phonon spectrum itself. In Fig.13 we show results for $\lambda_{\underline{k}} \equiv \lambda(\theta,\phi)$ along three constant ϕ arcs on the irreducible one fourty eight part of the aluminium Fermi surface. The dots are for $\phi = 0°$, the x's for $\phi = 22\ 1/2°$ and the open circles for $\phi = 45°$. We see variations in $\lambda(\theta,\phi)$ of the order of 25% which is certainly significant.

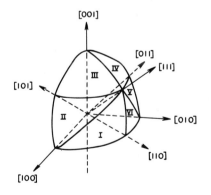

Fig.11:
The 1/48'th symmetry sections for the positive octant

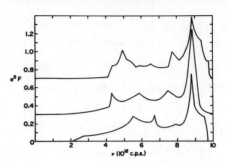

Fig.12: Free electron results for the directional spectral densities $\alpha_k^2(\omega)F_k(\omega)$ in Al. The lower curve is for $\underline{k}$ along the <100> direction the middle curve for <110> and the top curve for <111>.

5. Anisotropic Gap Equations

The Eliashberg gap equations in their four dimensional form which is appropriate for a pure single crystal anisotropic superconductor deal with a momentum and frequency dependent gap $\Delta(\underline{k},\omega)$ and

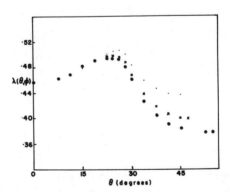

Fig.13: The directional electron-phonon mass enhancement parameter $\lambda(\theta,\phi)$ for Al calculated in the free electron model. The results are for three constant ϕ arcs namely $\phi = 0°$, $\phi = 22\ 1/2°$ (x) and $\phi = 45°$ (⊙).

renormalization function $Z(\underline{k},\omega)$ and are:

$$\Delta(\underline{k},\omega)Z(\underline{k},\omega) = \int_0^{\omega_c} d\omega' \frac{1}{(2\pi)^3} \int \frac{dS_{\underline{k}'}}{\hbar|V_{\underline{k}'}|} \text{Re}\left\{\frac{\Delta(\underline{k}',\omega')}{\sqrt{\omega'^2-\Delta^2(\underline{k}',\omega')}}\right\}[$$

$$\int_0^\infty d\nu \sum_\lambda |g_{\underline{k}'\underline{k}\lambda}|^2 \delta(\nu-\omega_\lambda(\underline{k}'-\underline{k})) K_+(\omega,\omega',\nu) - U_c] \qquad (18)$$

and

$$[1-Z(\underline{k},\omega)]\omega = \int_0^{\omega_c} d\omega' \frac{1}{(2\pi)^3} \int \frac{dS_{\underline{k}'}}{\hbar|V_{\underline{k}'}|} \text{Re}\left\{\frac{\omega'}{\sqrt{\omega'^2-\Delta^2(\underline{k}',\omega')}}\right\}[$$

$$\int_0^\infty d\nu \sum_\lambda |g_{\underline{k}'\underline{k}\lambda}|^2 \delta(\nu-\omega_\lambda(\underline{k}'-\underline{k})) K_-(\omega,\omega',\nu)] \qquad (19)$$

with

$$K_\pm(\omega,\omega',\nu) = \frac{1}{\omega'+\omega+\nu+i0^+} \pm \frac{1}{\omega'-\omega+\nu-i0^+}. \qquad (20)$$

So far, it has not been possible to iterate these equations to convergence for realistic metal parameters. Instead following a suggestion of Bennett /16/ a single iteration is carried out based on the solution of the isotropic equations for the metal. Specifically the isotropic solutions $\Delta(\omega)$ and $Z(\omega)$ of (7) and (8) are inserted into the right hand side of (18) and (19) to obtain

$$\Delta(\underline{k},\omega)Z(\underline{k},\omega) = \int_0^{\omega_c} d\omega' \text{Re}\left\{\frac{\Delta(\omega')}{\sqrt{\omega'^2-\Delta^2(\omega')}}\right\}[K_+(\omega,\omega':\underline{k})-N(0)U_c] \qquad (21)$$

and

$$[1-Z(\underline{k},\omega)] = \int_0^{\omega_c} d\omega' \text{Re}\left\{\frac{\omega'}{\sqrt{\omega'^2-\Delta^2(\omega')}}\right\} K_-(\omega,\omega':\underline{k}) \qquad (22)$$

with

$$K_\pm(\omega,\omega':\underline{k}) = \int_0^\infty d\nu \, \alpha_{\underline{k}}^2(\nu) F_{\underline{k}}(\nu) \left\{\frac{1}{\omega'+\omega+\nu+i0^+} \pm \frac{1}{\omega'-\omega+\nu-i0^+}\right\}. \qquad (23)$$

In this form we see that the anisotropic spectral densities

$\alpha_k^2(\omega)F_k(\omega)$ enter directly in the gap anisotropy. Once $\Delta(\underline{k},\omega)$ is known the anisotropic gap edge $\Delta_0(\underline{k})$ is obtained from the equation

$$\text{Re } \Delta(\underline{k},\Delta_0(\underline{k})) = \Delta_0(\underline{k}). \tag{24}$$

The set of equations (21) and (22) apply to strong coupling systems such as Pb as well as to weak coupling materials like Al. In the weak coupling limit Leavens and I /15/ have managed to simplify these equations considerably and obtain an accurate, yet much simpler, set of equations for the gap edge $\Delta_0(\underline{k})$. They are

$$\Delta_0(\underline{k}) = \frac{1}{Z(\underline{k})} \left(\int_{\Delta_0}^{\omega_0} d\nu\, \alpha_k^2(\nu)F_k(\nu)K(\nu,\omega_0,\Delta_0) - N(0)U_c \log\left(\frac{2\omega_0}{\Delta_0}\right)\right)\Delta_0 \tag{25}$$

where

$$Z(\underline{k}) = 1 + \int_{\Delta_0}^{\omega_0} d\nu\, \alpha_k^2(\nu)F_k(\nu)\left\{ L(\nu,\omega_0,\Delta_0) + \frac{2}{\nu+\omega_0} \right\} \tag{26}$$

where ω_0 is the maximum phonon frequency in $F(\omega)$ and Δ_0 is the solution of the isotropic dirty limit version of the simplified equations (25) and (26), namely

$$\Delta_0 = \frac{1}{Z} \left\{ \int_{\Delta_0}^{\omega_0} d\nu\, \alpha^2(\nu)F(\nu)K(\nu,\omega_0,\Delta_0) - N(0)U_c \log\left(\frac{2\omega_0}{\Delta_0}\right) \right\} \Delta_0 \tag{27}$$

with

$$Z = 1 + \int_{\Delta_0}^{\omega_0} d\nu\, \alpha^2(\nu)F(\nu)\left\{ L(\nu,\omega_0,\Delta_0) + \frac{2}{\nu+\omega_0} \right\}. \tag{28}$$

The kernels $K(\nu,\omega_0,\Delta_0)$ and $L(\nu,\omega_0,\Delta_0)$ are given in terms of simple and explicit combinations of logarithms and square roots in the variables ν, ω_0 and Δ_0. The functional form is specified in the paper by Leavens and Carbotte /15/ and does not need to be reproduced here.

In Fig.14 we show results for the Fermi surface variation of the gap edge in Al. The calculations are based on the set of equa-

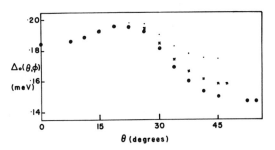

Fig.14: Fermi surface variation of the energy gap $\Delta_0(\theta,\phi)$ for Al in the free electron model. The results are for three constant ϕ arcs, namely $\phi = 0°(\cdot)$, $\phi = 22\ 1/2°$ (x) and $\phi = 45°$ (○).

tions (25) and (26) and the directional spectral weights of Fig.12. We stress again that band structure corrections have been left out of these distributions. We have taken the Fermi surface to be a sphere and the electronic wave functions to be plane waves. On comparison with the results for $\lambda(\theta,\phi)$ presented in Fig.13 we note that the gap edge $\Delta_0(\theta,\phi)$ varies in very much the same way. To make contact with BCS theory we define an anisotropy parameter $a_{\underline{k}}$ which measures the variation in $\Delta_0(\underline{k})$ from the average gap value by

$$\Delta_0(\underline{k}) = (1 + a_{\underline{k}})<\Delta_0(\underline{k})>. \tag{29}$$

The mean square anisotropy $a^2 = <a_{\underline{k}}^2>$ is given by

$$a^2 = \frac{<\Delta_0(\underline{k})^2> - <\Delta_0(\underline{k})>^2}{<\Delta_0(\underline{k})>^2} \tag{30}$$

and is found to be equal to $a^2 = .0084$ which is small but significant. We conclude that phonon anisotropy and Umklapp anisotropy together provide a significant source of variation in $\Delta_0(\theta,\phi)$ as well as in $\lambda(\theta,\phi)$.

6. Effects of Band Structure

We now want to discuss the modification in $\lambda_{\underline{k}}$ and $\Delta_0(\underline{k})$ that results from the introduction of band structure effects into the calculations. As I have already stressed it is important to realize that these complications enter the formalism only through modifications in the directional spectral densities. The mathematical structure of the two equations (21) and (22) remains unchanged. The real Fermi surface, the Fermi velocities and the multiple plane wave character of the electronic states come into (14) and (13). Four plane wave calculations of these quantities have been performed by Leung /17/ and results for $\alpha_{\underline{k}}^2(\omega)F_{\underline{k}}(\omega)$ for $\underline{k}$ along $\phi = 1°$, $\theta = 1°$ and $\phi = 23°$, $\theta = 15°$ are given in Fig.15 (solid curves). For comparison we also show on the same figure a directional frequency distribution

$$F_{\underline{k}}(\omega) = \frac{1}{(2\pi)^3} \sum_\lambda \int \frac{dS_{\underline{k}'}}{\hbar|V_{\underline{k}'}|} \delta(\omega - \omega_\lambda(\underline{k}' - \underline{k})). \tag{31}$$

This function is used in our numerical work for normalization purposes. It gives a measure of the anisotropy when the electron phonon coupling $g_{\underline{k}'\underline{k}\lambda}$ is set equal to 1 in formula (14). The differences in $F_{\underline{k}}(\omega)$ as a function of direction are more striking than those in $\alpha_{\underline{k}}^2(\omega)F_{\underline{k}}(\omega)$ because $g_{\underline{k}'\underline{k}\lambda}$ in (14) tends to reduce considerably the contribution from the large peak in $F_{\underline{k}}(\omega)$ around 5×10^{12} cps. Still,

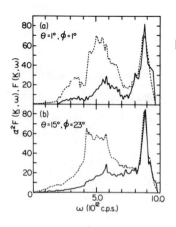

Fig.15: The directional spectral densities $\alpha_{\underline{k}}^2(\omega)F_{\underline{k}}(\omega)$ (solid lines) for Al including band structure effects. Also shown for comparison (dotted lines) is the function $F_{\underline{k}}(\omega)$. Figure (a) is for $\phi = 1°$, $\theta = 1°$ and figure (b) is for $\phi = 23°$, $\theta = 15°$.

differences with changing direction are still evident in $\alpha_k^2(\omega)F_k(\omega)$ which, of course, lead directly to differences in $\Delta_{\underline{k}}$ and $\lambda_{\underline{k}}$ as a function of position on the Fermi surface.

Before presenting results for $\lambda_{\underline{k}}$ and $\Delta_{\underline{k}}$ it is useful to make reference to Fig.16 in which we show a projection of the Al Fermi surface onto a sphere in the 1/48'th irreducible symmetry sector. The shaded regions indicate places at which there is no Fermi surface. This is most easily understood with reference to Fig.7 where the cross-section of the Al Fermi surface in the (001) plane is shown. If we introduce an angle ψ in this plane measured from the line ΓK and increasing towards ΓX we see that in the region of ψ around ΓW there is no Fermi surface for some ψ values.

The x's in Fig.16 indicate those points at which we have made calculations of the directional spectral densities. In Fig.17 we show results for the variation of $\lambda_{\underline{k}}$ along three constant ϕ arcs only. The most striking difference between these results and the results of Fig.14 which were obtained in the free electron model is that there are now regions (shaded regions) of θ where no Fermi

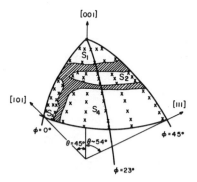

Fig.16: Schematic representation of the Al Fermi surface. The x's indicate the points at which we have carried out calculations. The shaded parts are regions in which there is no Fermi surface.

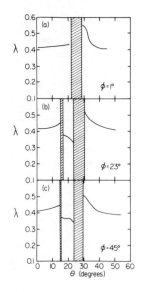

Fig.17: The Fermi surface variation of $\lambda(\theta,\phi)$ for Al when band structure effects are included. Figure (a) is for the $\phi = 1°$ arc (b) for $\phi = 23°$ and (c) for $\phi = 45°$.

surface exists. These regions fall near Bragg planes where the Fermi surface can be quite distorted and the multiple plane wave nature of the electronic wave functions becomes important. Discontinuities in the magnitude of $\lambda_{\underline{k}}$ are clearly apparent. The general trend is, however, roughly as in the free electron model.

We can introduce an anisotropy parameter $b_{\underline{k}}$ according to

$$\lambda_{\underline{k}} = (1 + b_{\underline{k}}) <\lambda_{\underline{k}}> \qquad (32)$$

and take $<b_{\underline{k}}^2> \equiv b^2$ as an overall measure of the anisotropy in $\lambda_{\underline{k}}$. We obtain

$$b^2 = .01 \qquad (33)$$

which is quite large. This value will shortly be compared with the anisotropy in the gap edge.

Our results for the gap edge /18/, fully accounting for Fermi surface anisotropy and the multiple plane wave character of the wave functions, are given in Fig.18. On comparison with Fig.17 we note a great deal of similarity between the variation of $\Delta_0(\underline{k})$ and

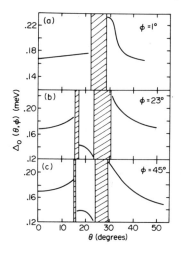

Fig.18: The Fermi surface variation of $\Delta(\theta,\phi)$ - the gap edge - in Al when band structure effects are included.

the mass enhancement $\lambda_{\underline{k}}$. The mean square anisotropy in $\Delta_0(\underline{k})$ is

$$a^2 = .0187$$

which is quite a bit larger than the anisotropy found for $\lambda_{\underline{k}}$. Also this value is more than twice the result found in the free electron model which includes only phonon and Umklapp anisotropy. We conclude that band structure effects make a very significant contribution to gap anisotropy in Al.

7. Average Gap and Critical Temperature in a Pure Crystal of Al

Based on the simplified set of equations (25) and (26) for $\Delta_0(\underline{k})$ Leavens and Carbotte /15/ were able to derive an approximate but analytic expression for the average gap in a pure single crystal weak coupling superconductor. It is

$$<\Delta_0(\underline{k})> = 2\omega_0 \times \exp \left\{ -\frac{1+(1+\frac{1}{2}<a_{\underline{k}}^2> + 2<a_{\underline{k}}b_{\underline{k}}>)\lambda+(1+<a_{\underline{k}}\bar{b}_{\underline{k}}>)\bar{\lambda}-\frac{1}{2}<a_{\underline{k}}^2>\mu^*}{(1+<a_{\underline{k}}b_{\underline{k}}>)\lambda-\mu^*} \right\} \quad (34)$$

where $\bar{b}_{\underline{k}}$ is similar to $b_{\underline{k}}$ and is given by

$$\bar{\lambda}_{\underline{k}} = (1 + \bar{b}_{\underline{k}}) < \bar{\lambda}_{\underline{k}} > \qquad (35)$$

with

$$\bar{\lambda}_{\underline{k}} = 2 \int_0^{\omega_0} \frac{d\nu}{\nu} a_{\underline{k}}^2(\nu) F_{\underline{k}}(\nu) (\log\{1 + \frac{\omega_0}{\nu}\}). \qquad (36)$$

The averages needed in formula (35) are

$$<a_{\underline{k}}^2> = .0187$$

$$<a_{\underline{k}} b_{\underline{k}}> = .0145$$

$$<a_{\underline{k}} \bar{b}_{\underline{k}}> = .0131$$

which leads to a ratio for the average pure single crystal gap to the dirty limit gap Δ_0 (obtained from (34) with the anisotropy set equal to zero) of

$$\frac{<\Delta_0(\underline{k})>}{\Delta_0} = 1.059. \qquad (37)$$

This is to be compared with a value of 1.022 in the free electron model. We note a six percent increase in the average gap over the dirty limit case. Also we note that band structure corrections, i.e. the distortions of the real Fermi surface from a sphere and the multiple plane wave character of the electronic wave functions, account for about half the increase in the average gap. The remainder comes from Umklapp and phonon anisotropy.

More interesting is the pure single crystal critical temperature T_c^{PSC}. It is given by

$$k_B T_c^{PSC} = 1.14 \, \omega_0 \, \exp \{ - \frac{1 + (1 + <a_{\underline{k}} b_{\underline{k}}>)\lambda + (1 + <a_{\underline{k}} \bar{b}_{\underline{k}}>)\bar{\lambda}}{(1 + <a_{\underline{k}} b_{\underline{k}}>)\lambda - \mu^*} \} \qquad (38)$$

where k_B is Boltzmann's constant. The ratio of T_c^{PSC} to T_c in the dirty crystal is

$$\frac{T_c^{PSC}}{T_c} = 1.091 \qquad (39)$$

which is to be compared with a value of 1.039 in the free electron model and of about 1.051 from experiments /15/. The order of magnitude agreement is good. Theory predicts a 9% increase in T_c as compared to a 5% increase found experimentally /15/. It should be kept in mind while doing this comparison that iteration of the complete 4 dimensional gap equations instead of using Bennett's first iteration result is likely to reduce somewhat the theoretical value of the anisotropy.

8. Results for the Case of Pb

Pb is the classic example of a strong coupling superconductor. In this case it is necessary to use equations (18) and (19) in order to calculate gap anisotropy as the gap function has important frequency dependence due to the fact that retardation effects are now significant. To calculate the right hand side of these equations for $\Delta(\underline{k},\omega)$ and $Z(\underline{k},\omega)$ it is necessary to know, in addition to the directional spectral densities $\alpha_{\underline{k}}^2(\omega)F_{\underline{k}}(\omega)$, the dirty limit gap $\Delta(\omega)$. This is obtained from a solution of the isotropic Eliashberg gap equations /14/ (7) and (8) based on the isotropic spectral weight $\alpha^2(\omega)F(\omega)$ for Pb. Our results for this quantity were presented in Fig.8 where they were also compared with tunneling results. Based on our results for $\alpha^2(\omega)F(\omega)$ we obtained the gap $\Delta(\omega)$ shown in Fig.19. What is shown is the real (solid curve) and imaginary (dotted curve) part of $\Delta(\omega)$ as a function of frequency. The large amount of structure in these curves is striking and reflects structure in $\alpha^2(\omega)F(\omega)$.

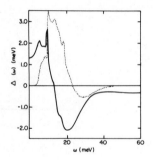

Fig.19:

Real (solid curve) and imaginary (dotted curve) parts of the isotropic frequency dependent gap $\Delta(\omega)$ for Pb based on our theoretical results for $\alpha^2(\omega)F(\omega)$.

Inserting the results of Fig.19 into (18) and (19) as well as our results for $\alpha_k^2(\omega)F_k(\omega)$ in Pb we obtain directional gaps $\Delta(\underline{k},\omega)$ as shown in Fig.20 for two directions of $\underline{k}$, namely <001> and <110>. At low frequencies significant differences are clearly seen between these two cases which lead directly to anisotropy in the gap edge which can be obtained from this information through solution of the equation

$$\text{Re } \Delta(\underline{k},\Delta_0(\underline{k})) = \Delta_0(\underline{k}).$$

The Fermi surface variation of $\Delta_0(\underline{k})$ is given in Fig.21. The solid dots represent points at which we have performed explicit calculations. The solid lines through the points were obtained by interpo-

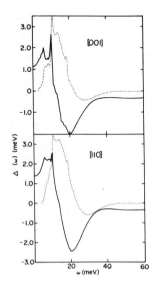

Fig.20:

Real (solid curve) and imaginary (dotted curve) parts of the anisotropic frequency dependent gap $\Delta_k(\omega)$ for Pb. The top curves are for $\underline{k}$ in the <001> direction while the bottom curves are for the <110> direction.

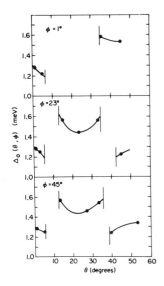

Fig.21: The Fermi surface variation of the gap edge $\Delta_0(\theta,\phi)$ in Pb.

lation of the calculated points. The value of the mean square anisotropy for Pb is $a^2 = .009$. This is smaller than for Al. No detailed comparison with experiment has been attempted so far. In view of the paper by Bostock and MacVicar /2/ this may be premature.

Our calculations are certainly only as good as our present knowledge of the input parameters, namely the details of the Fermi surface, electron-ion pseudopotential, Fermi velocities, electronic wave functions and lattice dynamics. While some uncertainties in these quantities certainly remain we believe that a possible source of error which is worth investigating is the use of the Bennett one iteration procedure. It is however extremely difficult in practice to remove this approximation since it would require making many iterations of the four dimensional Eliashberg equations. This would mean that we need to know, at each stage of the iteration, the gap at many points on the Fermi surface so that the small differences in $\Delta(\underline{k},\omega)$ from iteration to iteration are not lost in the computer noise.

References

/1/ J.R.Clem, Annals of Physics 40, 268 (1966)
/2/ J.L.Bostock, M.L.A.MacVicar, paper R-6, this volume, p. 213
/3/ G.M.Eliashberg, Soviet Phys. JETP 11, 696 (1960)
/4/ A.B.Migdal, Soviet Phys. JETP 7, 996 (1958)
/5/ Y.Nambu, Phys.Rev. 117, 648 (1960)
/6/ D.J.Scalapino, in Superconductivity, edited by R.D.Parks
    (Marcel Dekker, Inc., New York, 1969)
/7/ W.L.McMillan, J.M.Rowell, in Superconductivity, edited by
    R.D.Parks (Marcel Dekker, Inc., New York, 1969)
/8/ J.P.Carbotte, R.C.Dynes, Phys.Rev. 172, 476 (1968)
/9/ N.W.Ashcroft, Phil.Mag. 8, 2055 (1963)
/10/ H.K.Leung, J.P.Carbotte, D.W.Taylor, C.R.Leavens, Can.J.Phys.
     (in press 1976)
/11/ J.R.Anderson, A.V.Gold, Phys.Rev. 139, A1459 (1963)
/12/ M.Appapillia, A.R.Williams, J.Phys. F 3, 759
     (1973)
/13/ E.R.Cowley, Solid State Commun. 14, 587 (1974)
/14/ P.Tomlinson, J.P.Carbotte, Phys.Rev. B13, 4738 (1976)
/15/ C.R.Leavens, J.P.Carbotte, Annals of Phys. 70, 338 (1972)
/16/ A.J.Bennett, Phys.Rev. 140A, 1902 (1965)
/17/ H.K.Leung, Ph.D. Thesis, McMaster University (1974 unpublished)
/18/ H.K.Leung, J.P.Carbotte, C.R.Leavens, J.Low Temp.Phys. (in
     press 1976)

R-6 AN EVALUATION OF THE VALIDITY OF SUPERCONDUCTING EVIDENCE FOR ANISOTROPY AND MULTIPLE ENERGY GAPS

J.L.Bostock, M.L.A. MacVicar

Department of Physics, Massachusetts Institute of Technology, Cambridge, Massachusetts 02139, U.S.A.

Abstract

Implicit in the BCS theory of superconductivity is the possibility of anisotropic or even multiple energy gaps. Systematic deviations from the quantitative predictions of this theory were thus often ascribed to variations in gap parameter. Gap anisotropy models explain well the results of individual experiments to determine the thermodynamic properties of pure materials. Unfortunately, the anisotropies deduced from such experiments for any given material are inconsistent with each other, agreeing only in the limit of no observed anisotropy. Improved experimental expertise has generally reduced the anisotropy of a given material deduced from a particular experiment in direct proportion to the purity of the sample investigated. Various experimental approaches used to determine the gap anisotropy are discussed with special emphasis on the more selective methods such as superconducting tunneling. In particular, a detailed examination of reported tunneling results will be presented since tunneling is the most direct method of determining energy gap values. A recently developed model of imperfect tunneling structures (to be presented at this meeting) suggests that previously reported tunneling results are not convincing evidence for the existence of gap anisotropy or multiple energy gaps in any material. This interpretation, plus the total lack of consistency between anisotropy values determined by different experimental techniques, indicates that the experimental case for the existence of anisotropy of gap values is tenuous at best.

1. Introduction

BCS theory explains remarkably well most of the measured properties of superconductors /1/. One of the main results in that theory is an equation for the size of a forbidden region of energies in the normal electron energy spectrum caused by the binding up of pairs of electrons of opposite spins and momenta:

$$2\Delta(\underline{k}) = 2 \sum_{\underline{k}'} V(\underline{k},\underline{k}') \frac{\Delta(\underline{k}')}{2E(\underline{k}')} \tanh\{\frac{E(\underline{k}')}{2 k_B T}\}$$

Here $\Delta(\underline{k})$ is the energy gap between the ground and first excited state in the superconductor; $E(\underline{k})$ is the minimum energy required to create a quasiparticle of momentum $\underline{k}$ defined relative to the crystal axes; $V(\underline{k},\underline{k}')$ is the effective pairing interaction between electrons and phonons; and T is the absolute temperature ($T \leq T_c$). For ease in computation BCS assumed a spherical Fermi surface and an electron-phonon pairing potential constant for all pairing energies:

$$V(\underline{k},\underline{k}') = \begin{cases} V \text{ for } |E(\underline{k})|, |E(\underline{k}')| \leq \hbar\omega_{Debye}, \\ 0 \text{ for all other energies.} \end{cases}$$

From their analytic expression for the energy gap $\Delta(k)$, equations for both the thermal and electromagnetic properties of superconductors were derived. Thus, BCS theory deals with an idealized, isotropic superconductor and ignores the possibility of intrinsically anisotropic properties that might be due, for example, to crystallinity or to the complex multiply-connected Fermi surfaces known to exist for real metals.

Even at the inception of the theory two types of small, but systematic, deviations from the quantitative predictions of BCS were noted /2/. For the first type (observed in Pb and Hg) a theory which considers electron damping and time retardation of the electron-phonon interaction was developed (the strong-coupling theory) to successfully explain the departures of properties of Pb and Hg from BCS behavior. Specifically, the very strong electron-phonon interaction in these materials causes very high values of the gap ratio $2\Delta/k_B T_c$ and a positive deviation of the critical magnetic field curve from parabolic behavior.

For the other type of deviations not explainable in terms of strong-coupling theory, the anisotropic nature of real materials was assumed to be the source of the effects /3/. From a theoretical point of view /4/, there are three different sources of anisotropy: that of the electron band structure, that of the phonon dispersion curves, and that of the electron-phonon interaction. These three effects, if small, can collectively be expressed through a momentum-dependent pairing interaction $V(\underline{k},\underline{k}')$.

Anderson introduced a starting point for the problem of determining energy gap anisotropy /5/ by considering the effects on the electron pairing process of electrons scattering from impurities. His theory of dirty superconductors predicts that, assuming the energy gap is anisotropic in pure materials, Δ will be isotropic when there are sufficient impurities to make the electron scattering frequency larger than the gap frequency. In this limit electrons are scattered to all points on the Fermi surface, resulting in a pairing interaction which is averaged over the whole Fermi surface. Hence, the energy gap anisotropy normally present in the pure material would be "washed out". Studies of the far infrared absorption in pure and impure tin by Richards /6/ were considered an experimental confirmation of Anderson's ideas on gap anisotropy. In Fig.1, Richards' data on power absorption shows that as impurities are added to pure tin samples, the gap edge sharpens due to the increased Fermi surface averaging.

Anderson's ideas were developed in some detail /7/ by Markowitz and Kadanoff (MK). They assumed a BCS model of a weak coupling superconductor in which anisotropy is included in terms of a factorizable pairing (or interaction) potential

$$V(\underline{k},\underline{k}') = \{1 + a(\underline{k})\} V_{BCS} \{1 + a(\underline{k}')\}$$

where $a(\underline{k})$ is the temperature independent anisotropy parameter depending only on the orientation of the electron's $\underline{k}$ vector with respect to the crystal lattice and V_{BCS} is the constant potential of the BCS theory. In addition, MK assumed that the average of $a(\underline{k})$ over the Fermi surface is zero. To first order, then, all deviations of thermodynamic parameters from the BCS predictions are expressible in terms of a mean squared anisotropy parameter $<a^2>$, which is the average over the Fermi surface of $a^2(\underline{k})$.

For dilute impurity additions to superconductors, the MK theory predicts that the electron mean free path is reduced to less than

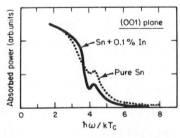

Fig.1: Frequency dependence of the absorbed far infrared power for the (001) plane of both pure and impure tin. The gap is defined as that energy where power absorption suddenly decreases, after /6/

the coherence length for electron pairs, producing less electron pairing and reducing the critical temperature for superconductivity. Further, the MK theory predicts a linear decrease in T_c with small concentrations of impurities (called the mean free path region). At larger concentrations, there are no longer simple independent scattering centers and chemical changes occur, altering the parameter V and causing dramatic changes in T_c (called the valence effect region). Figure 2 shows the general effects of impurities on T_c as predicted by theory and observed experimentally for a few isolated systems /7/.

Because of the simplicity of the MK approach, much of the gap anisotropy reported in the literature has been given in terms of $<a^2>$. To be a useful measure of anisotropy MK theory requires that on adding dilute impurities there be 1) no essential changes in electronic or phonon structure from those in the pure state, 2) no spin fluctuation effects introduced by the additions, and 3) that there be an effective homogeneity in, at least, local regions of space so that the impurity acts solely as an isotropic scattering center.

Clem, using the factorizable pairing potential of MK, derived /8/ the effects of anisotropy on the thermodynamic parameters of

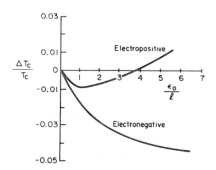

Fig.2: Experimental dependence of the shift in critical temperature as a function of mean free path (ℓ). ξ_0 is the superconducting coherence length. Electropositive impurities have less free electrons than the host; electronegative, more free electrons (after ref. /7/)

single crystal superconductors in terms of $\langle a^2 \rangle$. His expressions are those used by most experimentalists to classify their data and are the basis of most of today's discussion. Clem writes the mean squared anisotropy /9/ as:

$$\langle a^2 \rangle = \frac{\langle (\Delta_k - \langle \Delta_k \rangle)^2 \rangle}{\langle \Delta_k \rangle^2},$$

where Δ_k is an individual anisotropic energy gap value for electrons with momentum $\underline{k}$ defined relative to the crystal lattice and $\langle \Delta_k \rangle$ is an angular average of all the existing gaps Δ_k. The associated "directional-gap" equation

$$\Delta(\underline{k}) = \langle \Delta(\underline{k}) \rangle \{1 + a(\underline{k})\}$$

which maps the gap value onto the crystal lattice vectors has a temperature dependence that departs from the BCS gap dependence for $0.3 \lesssim T/T_c \lesssim 0.9$, where it falls measurably below BCS values. Clem also generalized this idea of a single energy gap intrinsically anisotropic in k-space, to the existence of multiple gaps in a given $\underline{k}$-direction, each varying in values as a function of crystal orientation. The analytical expressions for various superconducting properties he derives assume a uniform distribution of gaps from some minimum value to a maximum value determined by the best fit of experimental data to theoretical equations. The minimum Δ and maximum Δ define the two gap model of anisotropy.

Clem distinguishes between two different aspects of energy gap anisotropy according to experimental techniques: those from which only the existence of anisotropy can be deduced, and those which can be identified with selective processes that allow $\Delta(\underline{k})$ to be directly connected to various $\underline{k}$-directions in the crystal. The first group /10/ consists of macroscopic properties such as the high temperature specific heat, the change in T_c due to non-magnetic impurities, and the deviation of the critical magnetic field from parabolic behavior. These properties are determined by an angular average of Δ over all states on the Fermi surface, are directly proportional to $<a^2>$, and correspond to measuring a variation on the order of a few percent in experimental properties. Selective processes are those in which only certain groups of quasi-particles can be scattered on the Fermi surface so that these states can then be determined by varying the crystal orientation. Examples of such processes are superconductive tunneling, ultrasonic attenuation, microwave absorption, thermal conductivity, and the low temperature specific heat.

From comparisons of observed data to Clem's calculations, experimentalists have inferred the existence of an intrinsic gap anisotropy in many materials. In no case, however, has the comparison with theory been complete. To illustrate the significance of this remark, a brief review of various experimental means of determining energy gap anisotropy mentioned above will be presented and discrepancies between data and theory noted /11,12,13/. Since superconductive tunneling is in principle the most powerful selective technique for determining the actual values of anisotropic energy gaps as a function of crystalline orientation, a particular emphasis will be given to this experimental literature. In the discussion that follows, constant reference to experimental results for the transition metal niobium will be made to illustrate the importance of understanding the anomalies frequently observed in various experiments. The case of Nb is, in general, an instructive one since as a function of time, different experimentalists using totally different methods of determining gap anisotropy have independently come to the conclusion that there is no intrinsic anisotropy in Nb. In the course of this evolution quite a few assumptions made implicity in the standard interpretation of the various experimental observations (e.g. specific heat, thermal conductivity, and tunneling) have been shown to be false and many pitfalls of sample characterization recognized.

2. Specific Heat

Measurements of the electronic specific heat should, in principle, provide at least three characterizing parameters for anisotropy: the isotropic energy gap, the minimum available energy gap, and $<a^2>$. Clem showed that although the specific heat near T_c is itself insensitive to gap anisotropy because of thermal excitations, the magnitude of the jump is directly related to both $<a^2>$ and the value of the average energy gap (see Fig.3). As temperature decreases, the electronic specific heat in anisotropic materials is characterized by an increasingly smaller gap because of decreasing thermal excitations. The result, shown in Fig.4, is that with increased anisotropy there is a large and measurable excess specific heat. For real materials one expects an upward curvature in the plot of $\ln(C_{es}/T_c)$ vs T_c/T as temperature is reduced and the larger gap contributions cease to dominate. The very low specific heat contribution of the minimum energy gap (for $T/T_c \lesssim 0.3$) should then be exponential.

Unfortunately, a number of other factors also cause excess specific heats at low temperature: trapped flux, certain impurities, and strains due to dislocations in the sample /12,14/. Thus, a mapping

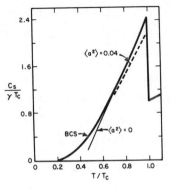

Fig.3:
The high temperature dependence of the specific heat near T_c, valid for $T/T_c > 0.5$ (after ref. /8/)

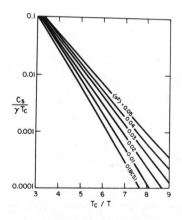

Fig.4:
The low temperature dependence of the specific heat for various $\langle a^2 \rangle$. Rectangular distributions of energy gap values are assumed for each curve such that $a_{min} = a_{max} = -\sqrt{3} \langle a^2 \rangle$. (After ref. /8/)

of the correct purity dependence of the electronic specific heat is an essential condition for associating the excess values of C_{es} to anisotropy. In the early studies of Ga, for example, no variation in low temperature slope as a function of purity was observed, precluding any conclusions on the intrinsic anisotropy of Ga. On the other hand, in a very recent study /12b/ of Al, the purity dependence of C_{es} was explored and complete agreement with the MK-Clem theory was observed.

In this same work /12b/ the behavior of zinc and cadmium were carefully studied. Preliminary experiments had indicated an extremely large gap anisotropy for Zn /12a/. In this most recent work, however, it was found that in the very low temperature region the specific heat was not exponential and the data could not be fitted even with Clem's two gap model. (At the very least, a three gap model would be required to fit the data /12b/.) Without a more definite study, then, there can be no conclusions on $\langle a^2 \rangle$.

The presence of exceedingly small amounts of impurities can severely affect data interpretations. Low temperature specific heat anomalies in Nb, previously thought to be due to either anisotropy, per se, or multiple energy gaps, are now thought to be due to the presence of small amounts of hydrogen /14/. The hydrogen, perhaps, enhances the naturally occuring resonant phonon scattering from dislocations and, thus, causes a radical change in phonon mean free path in the material. Only by making a full comparison of the experimental data to the MK-Clem theory, is there the possibility of identifying that such a problem exists. On the other hand, consistency of the comparison is not a guarantee that true anisotropy is being observed because indications that sample problems exist (i.e. giving incorrect purity dependence and/or differences in $<a^2>$ as determined by the specific heat jump at T_c and by the low temperature behavior) might be masked, as in the case of Nb, by strong-coupling effects. As Clem has pointed out /8/, for almost all the measurable thermodynamic parameters of superconductors, strong coupling effects, as compared to the effects of gap anisotropy, give corrections to the BCS theory which are opposite in sign. Hence, it is only for a known weak-coupling material that the MK-Clem theory can be applied unambiguously, and anisotropic behavior predicted.

3. Critical Magnetic Field Deviations

Clem also calculated the shape of the deviation of the critical magnetic field (CMF) curve from parabolic behavior for weakly coupled, but anisotropic, superconductors. For comparison with experiment, the CMF data is expressed in reduced variables as a deviation function:

$$D(t) = \frac{H_c(T)}{H_c(0)} - \{1 - (\frac{T}{T_c})^2\}$$

Both the true equilibrium thermodynamic critical field $H_c(0)$ and T_c must, however, be determined with extreme accuracy and precision if $<a^2>$ is to be extracted from experiment, since the shape of the deviation curve is extremely sensitive to their values. Because of this sensitivity, it is also essential that the purity dependence of $D(t)$ be determined to be BCS-like in the dirty limit. It is unusual to find a study wherein the purity dependence has been determined, but it is not unusual to find studies where there are discrepancies between the $<a^2>$ values determined from the low temperature and high temperature regions of the deviation curves /15/.

To explain the behavior of such CMF curves, particularly for materials known to exhibit other non-weak-coupling behavior, Gubser extended /16/ Clem's analysis to account for strong-coupling effects by defining an empirical coupling parameter, δ, which scales with the measured gap ratio $2\Delta(0)/k_B T_c$. (A δ of 1.00 is BCS-like and a $\delta \sim 1.20$ is strong coupling behavior.) For this extension of the theory, two independent measurements of thermodynamic parameters /16/ are then required to define $<a^2>$ (and δ). Typical CMF deviation curves are shown in Fig.5. The most negative deviation curve corresponds to Clem's calculation for $<a^2> = 0.04$; the positive deviation curve describes the very strong coupling materials such as lead and its alloys. In Table I values for δ and $<a^2>$ are given for a number of interesting materials. These values all fall well within an acceptable range for these parameters.

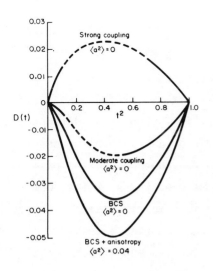

Fig.5:
Parameterization of deviations of the critical magnetic field curve from parabolic behavior as a function of anisotropy and coupling strength (after ref. /16/)

Table I: Anisotropy from critical magnetic field data (after D.U.Gubser /16/)

Metal	Coupling (δ)	$\langle a^2 \rangle$
Al	1.006	0.016
Cd	0.999	0.033
Ga	1.043	0.041
In	1.113	0.027
Ir*	-----	0.048
Mo	1.001	0.025
Pb	1.222	0
Sn	1.078	0.016
Ta	1.080	0.014
Th	1.041	0.019
Tl	1.079	0.041
Zn*	1.004	0.031

*Impurity problems.

4. Impurity Scattering

The MK theory of energy gap anisotropy explains the reduction of the superconducting critical temperature as a function of impurity addition /7/, i.e. the mean free path effect. Under the conditions previously discussed, the measured shift in T_c can be used to obtain the parameter $\lambda^i \langle a^2 \rangle$, where λ^i is a ratio of a transport process collision time to a characteristic time for superconducting pair breaking. As noted by the authors, the main weakness of the formalism is that the quantity λ^i is inaccessible to measurement. However, if the assumption that the impurity in the material acts as an isotropic scattering center holds exactly, then λ^i would in fact be unity. Thus, the ability of this theory to describe the intrinsic anisotropy of a material can in some sense be judged by the deviation of λ^i from unity. A second condition, of course, is that λ^i be the same for different kinds of impurities in a given host lattice.

In the initial experiments on Sn, the value of λ^i appeared to be the same for six different impurities /7/. However, this result may or may not be valid since the overall metallurgical characterization of the pure tin samples was incomplete. The results for Al and In, on the other hand required very different values of λ^i for consistency. Further experiments on Al, Sn, In, Zn, etc. ... /18/ all seem to indicate that each impurity addition has its own characteristic λ^i, which violates the theoretical assumption. In addition, for a few very careful experiments on various materials, the behavior of the shift in critical temperature in the dirty limit, is not that predicted /18/ by theory.

Thus, although impurity scattering as a method of determining anisotropy is, in principle, a good one, it is severely limited by strain effects, imperfect crystal structure, boundary scattering interference, and small but significant impurity addition changes to the host lattice.

A variation of this technique that eliminates the problems of chemical changes in the lattice is the determination of $<a^2>\lambda^i$ in pure superconductors solely by boundary scattering /19,20/. In this experiment the shift in critical temperature is determined as a function of sample thickness under the conditions that this thickness is much less than the bulk mean free path. Typical experimental results are shown in Fig.6, and the associated values of $<a^2>\lambda^i$ given in Table II. By comparing this data to other determinations of the anisotropy parameter $<a^2>$, a unique value of the scattering parameter λ^i can then be found. For the data given /13a/ in Table II, Al, In,

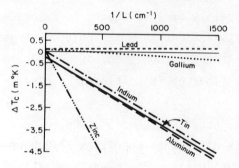

Fig.6: Shift in critical temperature of various elements from boundary scattering experiments. L is the sample thickness (after ref. /13a,20/)

and Ga would be characterized by $\lambda^i = 0.9$, which compare well to values of these parameters derived from impurity addition experiments.

Table II: Anisotropy from scattering effects in very pure materials (after W.D.Gregory and S.B.Horn /13a/)

Metal	$\lambda^i \langle a^2 \rangle$	$\langle a^2 \rangle$ (Gubser technique)
Al	0.011	0.016
In	0.018	0.027
Pb	< 0.001	0.000
Sn	0.014	0.016
Zn	0.029	0.031
Ga	0.005	0.008

Clearly, the boundary scattering experiment is one of the best for determining intrinsic anisotropy. The conditions under which it can be definitive are, however, very severe. The samples to be investigated must be free of inherent strain which is difficult to eliminate because of the sample geometry required; the scatter in the data must be extremely small; and finally, the metallurgical state of the pure, bulk reference sample must be complete. At the present state of the art, these conditions cannot be met simultaneously. In particular, the results presented in Fig.6 correspond to a very large scatter in data.

Finally, as noted by Gregory /19/, similar measurements on thin films of amorphous materials result in the same $1/\ell$ behavior as predicted for anisotropic superconductors. The mechanism behind the $1/\ell$ effect in amorphous films must be determined before the mean free path effect can unambiguously be attributed to the existence of intrinsic gap anisotropy.

5. Electromagnetic Radiation

Experimentally, energy gap anisotropy has been inferred from the temperature dependence /21/ of absorption of far infra-red radiation on single crystal faces, microwave absorption in bulk samples ($\nu \ll 2\Delta$), and bulk absorption of polarized radiation ($\nu \gtrsim 2\Delta$). Except for the very clear study /6/ by Richards on the behavior of pure and impure tin (Fig.1), the proper interpretation of much of this data is unclear /12,22/. For bulk absorption experiments the data seems to reveal only the "averaged" feature of anisotropy effects because of the complicated interactions within the given crystal structure. For most studies, a model for the anisotropic superconductor is assumed and the type of averaging is determined by how well the data agrees with the initial assumptions /22/.

In fact, very few studies have been done with single crystal samples; and, it is often reported that even when impurities are added to these samples, anisotropy effects such as multiple gap values, persist into the dirty limit where one would expect isotropic gap behavior. In surface impedance studies, again only "averaged" features of anisotropy can be observed because large angle scattering of electrons by the surface mixes the electron wavefunctions (and thus the gap values) from all different directions in the crystal. (This is, of course, true for all measurements in which the surface plays an important role such as boundary scattering and tunneling.)

6. Thermal Conductivity

Evaluation of anisotropy from thermal conductivity data is extremely difficult /12,23/. The standard data analysis techniques are complicated by the <u>details</u> of the scattering processes and the presence of a phonon component whose magnitude depends on the physical <u>perfection</u> of the sample. Changes in the electron mean free path in going from the normal to the superconducting state reproduce anisotropic gap behavior, as do the effects of residual strain in the material. The effects of impurities are peculiar to the material under examination and can reproduce anisotropic gap behavior. In fact, it appears to us that of all the methods for determining $<a^2>$, this method is probably the most sensitive to materials problems. Thus, if such a study results in $<a^2> = 0$, this is a definitive statement /24/.

Unlike some of the other techniques already discussed, the purity dependence of the thermal conductivity is not necessarily a reliable indicator of $<a^2>$ value primarily because of data analysis procedures /25/. Niobium is a case in point. In Fig.7 thermal conductivity data for both high and low resistivity Nb are shown. The relatively dirty sample has a behavior consistent with isotropic (or BCS-like) gap behavior; the high purity sample appears consistent with an anisotropic gap behavior. In fact, it has been shown /23,26/ that this behavior of the pure sample is due to the existence of a strong phonon scattering in the normal state rather than to increases of the gap value in the superconducting state. Similarly,

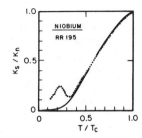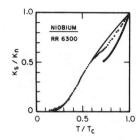

Fig.7: (a) Reduced thermal conductivity of a Nb sample with resistivity ratio 195. ——— theoretical calculation, ······· data (after ref. /23b/)

(b) Reduced thermal conductivity of a Nb sample of resistivity ratio 6300. ——— theoretical calculation, ······· data, ——— ratio of superconducting to normal state conductivity for phonon scattering for BOTH pure and impure Nb (after ref./23b/).

other anomalies in the low temperature thermal conductivity (and specific heat) previously observed in Nb correspond /23/ to changes in the phonon mean free path caused by small amounts of impurities which probably enhance the resonant dislocation scattering intrinsic to the material.

7. Ultrasonic Attenuation

Ultrasonic attenuation (UA) experiments are the most widely used for determining energy gap values of single crystal materials. The best and most careful studies of gap anisotropy using UA have been carried out with white tin. Typical data /27/ from such experiments are shown in Fig.8. As predicted by the theory of anisotropy, the attenuation systematically decreases as the purity increases; however, the curves for the pure samples (c and d) fall well below the appropriate BCS functions. In this particular study, the author argues convincingly that the temperature dependence as a function of impurity content /28/ can be explained in terms of the changes

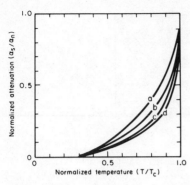

Fig.8:
The temperature variation of the normalized ultrasonic attenuation in single crystal tin: (a) heavily doped samples, propagation in <100> and <001> directions; (b) slightly doped samples, propagation in <100> and <001> directions; (c) pure sample, <001> propagation; and (d) pure sample, <100> propagation (after ref. /27/)

in electron and phonon mean free paths for a Clem-like two gap model of an anisotropic superconductor. Effective energy gap values for the various ultrasonic propagation directions in the basal plane of tin determined in a similar study /29/ are shown in Fig.9.

Unfortunately very few other materials have exhibited such clear and unequivocal data as that for tin probably because of complicated averaging effects similar to those in thermal conductivity and electromagnetic radiation experiments /12a,29,30,31/. In addition, there are some fairly sophisticated considerations of mean free path effects, as alluded to above, which can cause major deviations from BCS-like behavior and whose origin is not the anisotropy of the energy gap /30,32/. These latter effects are extremely important since they are largest at the onset of superconductivity which is one of the two critical regions for determining if gap anisotropy exists. The results of applying mean free path corrections to gap

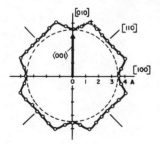

Fig.9:
Measured anisotropy of the zero temperature energy gap in pure tin. The arrow gives the value for propagation along <001>.
$A = 2\Delta(0)/k_B T_c$ and --- BCS value of $A = 3.52$; -o-o- and -x-x- measured gap values for propagation in the basal plane of the single crystal

An Evaluation of the Validity ... (Energy Gap)   231

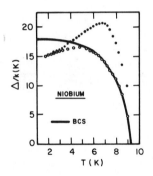

Fig.10:
The values of $\Delta(T)$ for a <110> Nb single crystal derived from ultrasonic attenuation data. —— BCS curve; ··· uncorrected values; ooo values obtained by assuming a change in electron mean free path on going from the normal to superconducting state due to thermal phonon scattering (after ref. /32/)

values determined by the UA technique are illustrated in Fig.10 /32, 33/.

It has also been shown quite recently that strain and dislocations can give rise to anomalous deviations from BCS behavior which are unconnected to gap anisotropy /34/. (These effects are different than those taken into account in standard UA data handling procedures /30/.) In single crystal tin, for example, corrections for such effects brings the experimental data into perfect coincidence with isotropic gap behavior /34/.

8. Summary of Results Presented in Sections 2 - 7

Before beginning a detailed discussion of the results of tunneling studies on single crystal substrates, let us consider the collective results concerning the existence of energy gap anisotropy obtained using the techniques previously mentioned. For any given material there are both bulk determinations and those which mainly probe the surface of the specimen. In general, the results of these two types of experiments yield very different results as can be seen from the data shown in Table III.

For Ga, which is typical of materials thought to be highly anisotropic, surface techniques result in extremely small values of $<a^2>$

Table III: Comparison of $\langle a^2 \rangle$ for Gallium (after W.D.Gregory /19/)

BULK		SURFACE	
Method	$\langle a^2 \rangle$	Method	$\langle a^2 \rangle$
Impurity Scattering	0.020	Boundary Scattering	0.0025
Specific Heat	0.023	Tunneling (1)	0.0025
Critical Magnetic Field	0.041	Tunneling (2)	$\lesssim 0.0020$
Ultrasonic Attenuation	0.075		
Thermal Conductivity	0.017		
Nuclear Spin Relaxation Time	0.011		

(~ 0.002), whereas bulk techniques give values from ~ 0.01 to 0.08 corresponding to a significant anisotropy in Δ. In this table, as in the one that follows, each individual experiment is subject to the limitations discussed earlier. Since the cause of the particular difficulties in a given experiment cannot be determined, the values of $\langle a^2 \rangle$ shown in Tables III and IV must stand until more definitive experiments can be performed or these results explained in terms of other phenomena /35/.

For other materials, listed in Table IV, there is not in general a large variation in the order of magnitude of $\langle a^2 \rangle$ values. In particular, Al and Pb seem to show fairly consistent behavior. Lead has no energy gap anisotropy and Al appears to have a large $\langle a^2 \rangle$ value on the order of 0.01 - 0.02. Still, in the case of Al as for the other elements listed in Table IV, a factor of two in anisotropy is well above the indeterminancy of any of the individual techniques used to determine $\langle a^2 \rangle$, and hence, the origin of the discrepancies experiment-to-experiment is not clear. In fact, the $\langle a^2 \rangle$ values reported for specific heat experiments correspond to Clem's distributed gap model of anisotropy, while those reported from ultrasonic data, for example, correspond to two individual gaps determined for each crystalline orientation, neither of which shows much k-space variation. Thus, in order to determine if these two experiments are consistent with one another, it is necessary to reproduce the characteristics of one experiment using the results of the other. As mentioned previously, such procedures rarely result in agreement between calculated and measured properties /12b/.

Table IV: Summary of Evidence for Gap Anisotropy in Superconductors (after K.Milkove /13b/)

Metal	Method	$<a^2>min$
Al	Electromagnetic Radiation	0.02
	Ultrasonic Attenuation	0.011 ± 0.002
	Ultrasonic Attenuation	0.02
	Specific Heat	0.010 ± 0.003
	Critical Magnetic Field	0.013
	Thermal Conductivity	None
	Impurity Scattering and Plastic Deformation	0.009
	Boundary Scattering	0.016
	Tunneling	0.009
	Tunneling	0.001
In	Electromagnetic Radiation	0.010
	Ultrasonic Attenuation	0.0003
	Ultrasonic Attenuation	0.010
	Specific Heat	0.010
	Critical Magnetic Field	0.027
	Thermal Conductivity	0.0008
	Impurity Scattering and Plastic Deformation	0.021
	Boundary Scattering	0.027

Table IV (cont.)

Metal	Method	$<a^2>_{min}$
Pb	Electromagnetic Radiation	None and 0.001
	Electromagnetic Radiation	0.0002
	Ultrasonic Attenuation	unstrained: none strained: at least 2 gaps
	Ultrasonic Attenuation	0.0009
	Specific Heat	0.33??
	Critical Magnetic Field	None
	Thermal Conductivity	None
	Impurity Scattering and Plastic Deformation	0.004
	Boundary Scattering	None
	Tunneling	0.007
	Tunneling	0.009
	Tunneling	0.031
Sn	Electromagnetic Radiation	0.020
	Electromagnetic Radiation	0.010
	Ultrasonic Attenuation	0.040
	Ultrasonic Attenuation	0.030
	Specific Heat	0.0035
	Critical Magnetic Field	0.022
	Thermal Conductivity	0.010
	Impurity Scattering and Plastic Deformation	0.023
	Boundary Scattering	0.016
	Tunneling	0.026

Table IV (cont.)

Metal	Method	$<a^2>_{min}$
Zn	Electromagnetic Radiation	0.009 or 0.033
	Ultrasonic Attenuation	0.003
	Ultrasonic Attenuation	0.003
	Ultrasonic Attenuation	0.040
	Specific Heat	0.068
	Specific Heat	0.020
	Critical Magnetic Field	0.031 ± 0.004
	Thermal Conductivity	0.030
	Impurity Scattering and Plastic Deformation	0.047
	Boundary Scattering	0.031
	Tunneling	None
Cd	Hypersound Attenuation	0.040
	Ultrasonic Attenuation	0.020
	Specific Heat	0.050
	Critical Magnetic Field	0.033
	Thermal Conductivity	0.010
	Tunneling	None

9. Tunneling

Tunneling experiments on single crystal substrates are considered the most direct method of determining the energy gap properties of superconductors, since the direction of tunneling electrons with respect to crystal axes can be carefully controlled. Both the variation of energy gap value with crystallographic orientation and the existence of multiple energy gap values along a given orientation are, again, commonly referred to as tunneling anisotropy. Todate, tunneling results reporting anisotropy have been interpreted either in terms of electronic band structure or the anisotropy of phonon characteristics. In the former, correlations of observed energy gap(s) with Fermi surface behavior are attempted; multiple gap values are assumed to arise from the various Fermi surface zones contributing electrons along a given tunneling direction. The relative heights of the current jumps at the gap edge energies in I-V curves for the junction are taken as indicators of the current contribution of electrons from each zone. For phonons, comparisons of $\frac{dI}{dV}$ - V and $\frac{d^2I}{dV^2}$ - V line shapes and measured critical energies are made to theoretically calculated line shapes and predicted critical points.

The constant theme running through the tunneling literature is that of inconsistency: inconsistency in gap value results between one experimenter and another, and even inconsistency between the earlier and the later results of a particular experimenter. The reason for this is that there are a number of problems involved in undertaking such experiments. The fabrication of both the electrodes and the barrier of junctions is difficult; the interpretation of characteristics is not agreed upon; it is not clear when an effect being observed is a surface or a bulk property; and finally, the actual selection rules operating for tunneling processes are not understood. In our laboratory, we have, over a long period of time, gained an understanding of tunneling behavior intrinsic to at least one material, single crystal Nb. We have also gained a greater appreciation for the pitfalls in general junction fabrication processes; a sense of what is a valid or an invalid approach to data analysis; and a skepticism concerning conventional interpretations of tunneling data.

The number of materials for which tunneling studies have reported anisotropy effects is less than ten and of these only Nb, Sn, Al, Pb and Ga have been studied extensively. In assessing the results of these and other tunneling studies, many questions must be kept in mind, such as: is the sample surface really representative of the bulk material ? Are strains, thermal facets, micro-faults, defects, impurities, contaminants, or inhomogeneities present? Is the tunneling barrier uniform, insulative, stable and amorphous? Are the metal-barrier interfaces clean? Is the counterelectrode reliably isotropic and single-valued in energy gap? What criteria are used to judge junction performance? How are gap values determined from the data? How reproducible is the data? Are anomalous structures present in the characteristics? What are the effects of temperature, aging, and magnetic field? Etc. ... Only with satisfying answers to such questions can confidence be felt in the interpretation of tunneling results in terms of fundamental gap behavior.

9.1 Niobium

In 1968, MacVicar and Rose reported the existence in single crystal niobium of a single, anisotropic energy gap /36/. Cylindrical single crystals were grown in uhv by the floating zone electron beam method to provide surfaces of high structural perfection. The niobium was oxidized outside the vacuum system in warm oxygen; counterelectrode stripes were then deposited as narrow stripes aligned parallel to the crystal axis. In Fig.11 it is seen that particular ranges of magnitude for the gap appeared to relate somewhat to Fermi surface features /37/. Confidence in reporting anisotropy rested on the state-of-the-art quality of the junctions, on the qualitative agreement of the results to reported ultrasonic studies on Nb, on the information available at the time concerning niobium oxidation kinetics and the nature of the counterelectrode, and on the reproducibility of the data. However there was still a long list of nagging questions and seemingly small, but puzzling, inconsistencies concerning the data.

One area of concern was the occasional occurence of junctions with tunneling anomalies in the characteristics, e.g. low bias subharmonic structures, a dip (or "knee") just above the sum peak bias

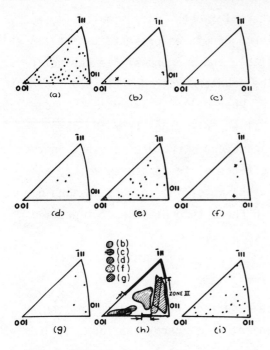

Fig.11:
(a) Stereographic plot of the Nb crystal orientations studied; (b) x denotes gap value of 2.82 ± 0.02, · denotes 2.98 ± 0.02, and ? denotes 2.94 ± 0.02 meV; (c) gap values of 3.02 ± 0.02 meV; (d) gap values of 3.06 ± 0.02 meV; (e) gap values of 3.10 ± 0.02 meV; (f) · denotes gap values of 3.14 ± 0.02; + denotes 3.18 ± 0.02 meV; (g) gap values of 3.22 ± 0.02 meV; (h) equi-gap value regions superimposed on approximated regions of Mattheiss's central section of niobium Fermi surface; (i) orientations exhibiting anomalous low bias gap-like structure (after ref. /37/)

(Fig.12), and a gap-like structure at low energy that appeared erratically with no crystallographic dependence (Fig.11i). Another area of concern was lack of agreement between analysis procedures and data criteria used in arriving at gap value determinations. In particular, Δ_{Nb} determined from multiparticle peak structure did not always agree with Δ_{Nb} determined from the sum peak structure; and it mattered whether one chose the maximum in the sum gap peak or the midpoint of the peak for $(\Delta_{Nb} + \Delta_{counterelectrode})$ in the sum peak energy gap determination. A further study of single crystals of rhenium grown similarly to niobium but utilizing an amorphous carbon layer as a tunneling barrier showed Δ_{Re} determined from some multiparticle structure, again, did not agree with the value determined from other related structure /38/. Even the sense of the anisotropy for gap values determined from the multiparticle structures differed from that determined by the sum peak.

Subsequently, we developed an ability to prepare nearly ideal

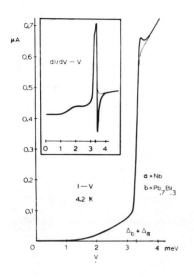

Fig.12:
I vs V characteristic and dI/dV-V for a single crystal Nb/Ox/Pb$_{0.7}$Bi$_{0.3}$-film junction exhibiting a negative resistance region (or "knee") just above the sum gap bias, $\Delta_{Nb} + \Delta_{PbBi}$. The solid line is the observed behavior; the dotted line is the behavior expected for a nearly-ideal junction

Nb junctions with characteristics free of subharmonic structure, knees, and anomalous structures as shown in Fig.13. Careful analysis

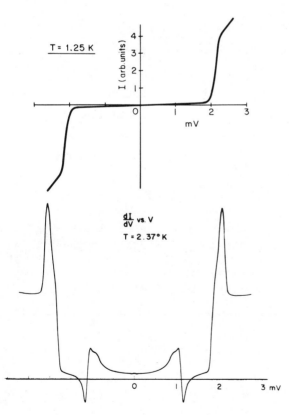

Fig.13: Typical I vs V and dI/dV-V characteristics for nearly-ideal single crystal Nb/Ox/In-film junctions (after ref. /39/)

of such data, with attention to internal consistency and to the development of criteria for judging data quality (not always the same as junction quality) led us to conclude there really is no anisotropy of the tunneling energy gap in niobium /39/.

9.2 Tin

Zavaritskii reported in 1964 the first systematic study of anisotropy in a single crystal superconductor /40/. He investigated Sn crystal surfaces which had been formed by contact with a glass mold; the glass was then broken off to expose the surface to atmosphere and oxygen. Considerable barrier fabrication difficulties were encountered, resulting in non-ideal tunneling characteristics. I-V and dI/dV-V data exhibited considerable nontunneling current contributions for resistances of 0.1 to 10 Ω.

Zavaritskii introduced the interpretation of multi-peaked derivative sum gap structure in terms of multiple energy gaps, and the assumption that the strength of derivative structure relates to different levels of current contribution by electrons from different parts of the Fermi surface. Distinct k-space boundaries to gap value regions were noted and equi-gap value surfaces were in some cases

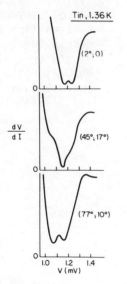

Fig.14: Illustration of Zavaritskii's dV/dI-V data for tin single crystals of three orientations. Note the sum gap structure at approximately 1.16 meV which is common to each curve (after ref. /40/)

not dissimilar to theoretical Fermi surface zones of tin. In Fig.14, dV/dI-V data for three crystal orientations is shown. One gap value is common to all three curves; multiple gap values correspond to the structures of approximately equal displacement in energy above and below this common structure. Typically, the author recorded multiple gap values only for cases where "substantial differences in values of 2Δ" were observed and not for cases of less resolved structure in the characteristics. Broad ranges of orientation often showed constancy of 2Δ to within 2%, while sometimes across only 5° ranges a 20% variation in 2Δ was reported.

In tin films (as opposed to single crystals) Zavaritskii observed only one, isotropic gap. However, Campbell and Walmsley reported that for thick enough films, two gaps are seen /41/. The sharp current jump at the gap edge in I-V for the thin film becomes broader for a thick film; the narrow, tapered sum gap peak in dI/dV-V for a thin film becomes double-peaked for a thick film. The authors assumed that boundary scattering washed out anisotropy effects in their thin film data. Anisotropy effects were expected in thick films, on the other hand, due to the tendency of films, in general, to have a preferred crystalline orientation /41/.

9.3 Aluminium

Similarly, Campbell and Walmsley saw two energy gaps in thick films of aluminium /41/. Not only did the derivative exhibit structure identifiable with two energy gaps, but the I-V clearly contained two current steps as shown in Fig.15a. In order to account for this characteristic, the authors adopted a procedure of synthesizing a matching theoretical I-V from separate I-V characteristics corresponding to quite different possible single gap values in the aluminium.

Wells, Jackson and Mitchell /42/ looked at both thin single crystal and thin polycrystalline aluminium films. These authors saw

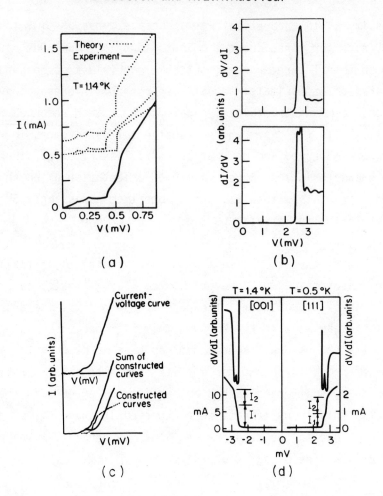

Fig.15: (a) Current voltage characteristic for tunneling into thick aluminium films at 1.14 K. Dashed curves illustrate Campbell and Walmsley's theoretical approach to synthesizing the experimental curve. (The origins of the theoretical curves are shifted vertically for clarity.) (After ref. /41/).

(b) Lykken et al. dI/dV-V curves for lead-insulator-lead tunnel junctions with a 3.5 μ-thick <110> lead film; (a) original junction; (b) second junction made with same single crystal film of lead after it had been reoxidized in pure O_2 for 2 hours at 150 °C (after ref. /44/).

(c) Possible multiple gap phenomenon reported by Gregory et al. for c-axis Ga-Ga_xO_y-Pb flat surface sample. Two theoret-

only one gap value, consistent with the thin film results of Campbell and Walmsley; however, the observed value was not the same in the two orientations investigated. From fairly limited data, Wells et al. concluded that anisotropy in $2\Delta_{Al}$ existed despite the fact that the two values were thought to be "too close to the isotropic energy gap" value in magnitude for a satisfactory anisotropy. In particular, the authors observed that changing either the substrate temperature during deposition, the substrate itself, or the counterelectrode greatly affected their measured gap. Stress effects were found to be large in the aluminium, significantly changing $2\Delta_{Al}$ and T_c.

Recently, Blackford /43/ reported successful tunneling into bulk single crystal aluminium. The derivative sum peaks were observed to be broader than those seen for thin aluminium films and their breadth varied with crystallographic orientation. Occasionally, multiple (or split) sum peaks were observed; the author omitted them from analysis because they were "considerably weaker than the main peak". However, he did interpret the variable breadth of the main peak as indicative of anisotropy. Multiple gap structure seemed to correlate to low impedance junctions. As the author himself noted, particularly worrisome is the fact that in one crystallographic direction investigated on two different junctions, the measured gap values differed by ten times the stated precision of the determination.

9.4 Lead

Figure 15b is tunneling data from single crystal lead reported

---

ically constructed curves are summed to synthesize a composite curve for comparison to the experimental current-voltage curve (after ref. /47/).

(d) Blackford's I-V and dV/dI-V curves for tunneling in the <001> and <111> directions, respectively, of a Pb single crystal. Note the difference in the ratio I_2/I_1 for the two cases (after ref. /45/).

by Lykken et al. /44/. Two gap values were seen along each of six directions in the crystals, with one high symmetry direction exhibiting a third gap. Resolution of complex gap structure depended on junction impedance, as seen in the change of appearance of the derivative sum peak as impedance increases. The authors found that the relative current step heights in the I-V curves of their samples were not consistent for similarly oriented junctions.

Blackford et al., however, found that the relative current jumps did vary with orientation in single crystal lead films /45/. Figure 15d shows I-V derivative data for two orientations investigated. The relative current step heights I_1 and I_2 were used as weighting functions for contributions to the tunneling current from various parts of the Fermi surface, with mixed results. Contrary to Lykken, Blackford et al. measured only two gaps in all directions investigated. Both were observed to be anisotropic, but the smaller value was more so than the larger value. In general, the gaps did not change value as much with orientation as they were different from one another at a given orientation. Blackford rested his case for the intrinsic nature of the two gaps he saw on a qualitative investigation of the derivative as a function of temperature.

Rochlin /46/ studied tunneling into thick polycrystalline lead films which exhibited considerable low bias, non-ideal structure. He deduced several critical values of energy which he felt corresponded to either energy gap values or to critical points of the E(k) surface in k-space, and reported at least three firm gap values; he also listed five additional critical points. Different combinations of these gaps and critical values were observed on different samples. Thermal cycling of samples caused the multiple gap-related structure to disappear; waiting a few days at room temperature and retesting brought the multiple gap structure back again. It was noted that no multiple gap structures were seen for any lead thickness when the counterelectrode was a normal metal. This was true even for an aluminium counterelectrode junction which did show multiple gap structure when the junction temperature was reduced below T_{cAl}. No anisotropy (or multiple gap) effects were seen in Pb-Pb junctions even for very thick films.

Campbell and Walmsley, like Rochlin, observed multiple energy gaps in thick lead films utilizing an aluminium counterelectrode /41/. However, Campbell and Walmsley could see multiple energy gap effects in a lead thickness four times less than that required by Rochlin. These authors never saw other than two gaps, except in very thin films, where they saw only one. From sample to sample the authors noted considerable quantitative variation in gap values.

It is worth remarking as regards lead that when Banks and Blackford /45/ fabricated lead films on various supporting substrates, they observed that the nature of the substrate significantly affected gap behavior. In particular, the single gap value observed for thin lead films varied beyond experimental error when the supporting substrate was changed. The authors concluded that the single crystal lead gap values originally reported were too small due to strain in the junctions, but that the original conclusions concerning the existence of gap anisotropy was still valid.

9.5 Gallium

Gregory and his co-workers /47/ have reported multiple gap values in single crystal gallium. In Fig.15c, I-V data for tunneling along the c-axis is shown, along with two I-V curves for different single gap values used to synthesize a composite I-V that matches the experimental curve. In one case, a fit was done with the unphysical gap values of $2\Delta = 1.6\, k_B T_c$ and $7.2\, k_B T_c$. The authors used only I-V plots to analyze their curves, believing that the current jump ratio is the primary gap indicator. The authors indicated concern for the quality of the I-V plots, however, and for the possibility of barrier imperfection due to their fabrication procedure /47/. Difficulties including nontunneling current, sample aging, and data interpretation were noted; junction impedances were very high.

Yoshihiro and Sasaki /48/ observed much the same variation of energy gap with Fermi surface location as did Gregory; however, they did not report multiple gaps. Their surface preparation differed quite markedly from that of Gregory et al. In particular, they subjected the oxidized barrier layer to ion bombardment in order to lower the impedance of the junctions. Polick did observe apparent multiple gap effects for single crystal gallium /49/, but did not feel these were intrinsic effects because nontunneling currents were common in his characteristics. This author felt that their presence, as well as the presence of subharmonic multiparticle structures, precluded any statements concerning anisotropy. He saw two sum peaks, as in Fig.16, when tunneling along the a-axis. In the b- and c-axes, however, single gap values were seen, contrary to Gregory's observation /47/. The shift in relative peak heights with temperature shown in the figure is particularly interesting in view of both Blackford's interpretation of the higher peak as the primary value characterizing $2\Delta_{Al}$ in his single crystal aluminium tunneling data /43/, and Zavaritskii's approach to the tin data analysis /40/.

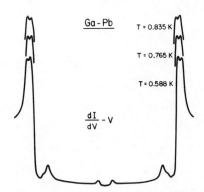

Fig.16: dI/dV-V as a function of temperature for a-axis single crystal gallium tunneling observed by Polick (after ref. /49/)

Gregory has recently observed tunneling into single crystal gallium junctions with a normal counterelectrode /50/. He has not seen the multiple energy gaps originally reported in Ga samples having a superconducting counterelectrode. This result is consistent with Rochlin's observations for lead /46/.

9.6 Phonons

A natural investigation that follows from results that seemingly indicate the existence of anisotropy is to look at the phonon structure in tunneling curves. It is expected that a set of longitudinal and transverse peaks should exist for each multiple gap value observed. In Blackford's single crystal lead film study, the shapes and amplitudes of observed phonon structure do seem to relate to the ratio of the current jumps in the I-V characteristic of a multi-gap junction /45/; however, only the transverse phonon structure showed such effects, the longitudinal structure did not. When Leger and Klein /51/ made tunneling experiments from Al into thick oriented lead films, they observed only a single Tomasch series and no phonon splitting despite seeing two lead gaps. In particular, no effects at the transverse peak were seen. Later, Blackford reported that transverse-peak splitting was indeed observed and that the degree of splitting matched the difference in magnitudes of the multiple gap values he measured in the direction of observation /45/. However, splitting did not occur in all samples but only in those where both a distinct separation of gap sum peaks was observed in the derivative and where the dominant current contribution corresponded to the smaller gap value. Also, the smaller gap value seemed to be the reference gap for the phonon structures. Splitting was more dependent on the degree of resolution of the two sum peaks than on the difference in gap values determined from the peaks. Interestingly, Banks and Blackford /45/ remarked that although transverse phonon effects were observed in lead films supported by single crystal lead substrates, they did not appear in lead film data when the films were deposited on glass slides.

9.7 Model

A summary of the five materials discussed is shown in Table V. The _same_ experimentalist sees _different_ results from thin films of a given material than from thick; also, _different_ experimentalists see different results from the _same_ kind of sample. The number of gaps observed is not consistent. The degree of anisotropy $<a^2>$ seen is variable.

Table V: Anisotropy from Tunneling

Metal	Format	Reference	No.Gaps	No.Directions	$<a^2>$
Al	thin films	41 (1967)	1		None
	thick films	41 (1967)	2	1 ?	0.129
	crystal	42 (1970)	2	2	0.001
	crystal	43 (1976)	1	24	0.009
Pb	thin films	41 (1967)	1		None
	thick films	41 (1967)	2	1 ?	0.002
	thick films	46 (1967)	3 (8)		0.008 (0.031)
	crystal	45 (1969)	2	10	0.007
	thick films	51 (1971)	2	1 ?	0.003
	crystal	44 (1971)	2 (3)	5	0.007 (0.009)
	thin films	45 (1973)	1		None
Sn	crystal	40 (1965)	Many	Many	0.026
	thin films	41 (1967)	1		None
	thick films	41 (1967)	2	1 ?	0.0007
Ga	crystal	48 (1970)	1	33 from <100> to <010>	0.001
	crystal	47 (1971)	2	22 from <100> to <010>	0.002
	crystal	49 (1973)	1	3 (axial)	None
	crystal	50 (1976)	1	15 from a axis to b axis	0.001
Nb	crystal	36 (1968)	1	51	0.004
	crystal	39 (1973)	1	35	None

In an effort to model such inconsistent behavior, without the concept of gap anisotropy, we have considered the existence of parallel tunnel current paths from one electrode to the other /52/. We will present the model in detail in the following paper. Briefly, the model assumes that a tunneling junction is represented by several possible current paths in parallel with one another between the two electrodes of the junction as shown in Fig.17. These parallel paths include an intrinsic obstacle in one or more paths. Physically, such an obstacle might correspond to a structural defect in the barrier or counterelectrode, a compositional homogeneity in either electrode, the existence of strain somewhere in the junction, the presence of uneven surface contamination, or some other non-ideal features typical of a "real world" junction. The model explains the following features of non-ideal tunneling /13b/.

(1) The existence of several erratically observed energy gap values along a particular orientation in a given superconductor, such as in the investigations on single crystals of aluminium, lead, and gallium.

(2) The observations in aluminium, lead, and tin of multiple gaps in thick films versus single gaps in thin films.

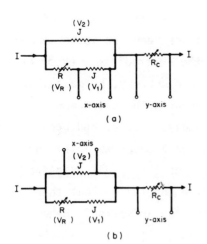

Fig.17:
Schematic representation of a tunneling junction with effectively two parallel current paths between its electrodes. R_c is a current-measuring resistor whose output is displayed on the y-axis of an x-y recorder; either voltage V_1 or V_2 is displayed on the x axis. R represents the presence of a non-ideal physical feature in the junction geometry causing a voltage drop V_R in one current path.

(a) Displaying V_1 results in an I-V characteristic with anomalous gap-like structure below the true sum gap bias.

(b) Displaying V_2 results in an I-V characteristic with anomalous structure above the true sum gap bias.

(3) The more anisotropic behavior of one of the two gaps observed in lead.

(4) The rough agreement of equi-gap value regions to main Fermi surface features for tin, lead and gallium.

(5) The validity of the ratio of sum gap current jumps in Al, Pb, Sn, and Ga I-V's as a measure of junction quality but not as an indicator of the relative current contributions of electrons from different zones.

(6) The incorrectness of associating relative sum-gap-structure strengths in dI/dV-V with relative current contributions of electrons from different zones as has been attempted for lead and tin.

(7) The sensitivities of the anisotropy results of all the tunneling studies discussed to fabrication conditions and thermal histories of samples, including impedance, oxidation history, electrode thickness, and presence of strain and aging.

(8) The inconsistent reports of splitting effects and nonsplitting of the phonon tunneling structure in thick lead films.

(9) The fact that the reported temperature dependences of observed energy gaps do not eliminate the possibility of non-intrinsic explanations of the data.

Thus, this model might provide a possible framework for understanding the various conflicting observations reported in the experimental tunneling literature. Because of the contradictory evidence concerning the existence of multiple gaps/anisotropy, because of the qualifying remarks often offered by different authors concerning peculiarities of the samples or data or analysis used, and because of the capability of our parallel current path model to unify the various reported observations irrespective of the actual existence of energy gap anisotropy, we are led to the conclusion that there is _really_ no convincing tunneling evidence for anisotropic behavior in any of the superconductors listed in Table V (or for the others we have not discussed here). Instead, a strong case emerges for the fact that junction fabrication and materials characterization considerations (especially strain) may underlie the interpretations of tunneling results in terms of anisotropy.

10. Conclusion

It is highly likely that the metallurgical state of the various materials investigated by the other experimental techniques already discussed also play an important role in determining the effects attributed to gap anisotropy. In these experiments, as in tunneling, there has been a general reluctance of investigators to pursue discrepancies between theory and various experimental results. Instead, there has been a general appeal to the collective results of all studies for the given material. The fairly universal belief in the existence of intrinsic anisotropy in superconductors is based on these collective results despite the fact that there are basic problems in interpretation and consistency for all of the various methods of determining anisotropy.

Certainly, systematic deviations from the laws of the BCS theory do exist and theories based on anisotropy and/or multiple energy gap(s) may be the best explanation of such deviations. But, the fact remains that there is no unambiguous evidence for the existence of an intrinsic anisotropy in the superconducting energy gap.

The authors wish to thank the National Science Foundation (MLAM) and the Massachusetts Institute of Technology (JB) for making it possible for them to travel to this meeting.
We thank the staff of the Atominstitut der Österreichischen Universitäten for their hospitality during our visit.

References

/1/ J.Bardeen, L.N.Cooper, J.R.Schrieffer, Phys.Rev. 108, 1175 (1975)
/2/ J.Bardeen, J.R.Schrieffer, Progress in Low Temp.Phys. III, (North-Holland Publ.Comp., Amsterdam 1961) p. 170
/3/ Other, more empirical methods of parameterizing observed data exist which are not based on anisotropic gap behavior. See, e.g. T.P.Sheahen, Phys.Rev. 149, 368, 370 (1966) J.Bostock et al., phys.stat.sol.(b) 62, 573 (1974); H.Padamsee et al., J.Low Temp.Phys. 12, 387 (1973)
/4/ V.L.Pokrovskii, Soviet Phys. JETP 13, 447 (1961); V.L.Pokrovskii, M.S.Ryvkin, Soviet Phys. JETP 16, 67 (1963); B.T.Geilikman, V.Z.Kresin, Soviet Phys. Sol.State 5, 2605 (1964); A.J.Bennett, Phys.Rev. 140A, 1902 (1965)
/5/ P.W.Anderson, Phys.Chem.Solids 11, 26 (1956)
/6/ P.L.Richards, Phys.Rev.Lett. 7, 412 (1961)
/7/ D.Markowitz, L.P.Kadanoff, Phys.Rev. 131, 563 (1963)
/8/ J.R.Clem, Ann.Physics 40, 268 (1966); Phys.Rev. 153, 449 (1967); Ph.D. Thesis, University of Illinois (Department of Physics), 1965, unpublished
/9/ A more fundamental explanation for $<a^2>$ will be discussed later in this conference by N.C.Cirillo and W.L.Clinton, paper C-15, this volume, p. 283
/10/ These thermodynamic quantities depend upon the probability for thermal excitation of quasi-particles so that for thermal energies ($k_B T$) comparable with the available gap values, the probability for thermal excitation is appreciable for all directions. Hence, it is essentially the angular average of the gap parameter that is being measured which, to first order, is proportional to $<a^2>$.
/11/ In a review of this nature, it is not possible to credit all authors for their work. The reader is referred to Refs. /12/ and /13/ for individual references.

/12/a A.G.Shepelev, Soviet Phys.Usp. 11, 690 (1969)
   b J.D.N.Cheeke, E.Ducla-Soares, J.Low Temp.Phys. 11, 687 (1973)
/13/a S.B.Horn, M.Sc.Thesis, Georgetown University (Department of Physics), 1973, unpublished
   b K.R.Milkove, S.M. Thesis, Massachusetts Institute of Technology (Department of Physics), in progress
/14/ V.Novotny, P.P.M.Meincke, J.Low Temp.Phys. 18, 147 (1975); G.J.Sellers, A.C.Anderson, Phys.Rev. B10, 2771 (1974)
/15/ In the Clem formulation the expressions for D(t) are valid only in the very high and very low temperature regions; and in all cases, $<a^2>$ must be small, $\lesssim 0.04$.
/16/ D.U.Gubser, Phys.Rev. B6, 827 (1972)
/17/ Although this extension to the Clem theory brings in an additional fitting parameter, the extreme sensitivity of the CMF deviation to even small strong coupling effects requires some means of accounting for them if this function is to be useful. See, also, Ref. /3/.
/18/ W.D.Gregory, Ph.D. Thesis, Massachusetts Institute of Technology (Department of Physics), 1966, unpublished; D.Farrell, J.G.Parks, B.R.Coles, Phys.Rev.Lett. 13, 328 (1964); G.Boato, G.Gallinaro, C.Rizzuto, Phys.Rev. 148, 353 (1966)
/19/ W.D.Gregory, M.A.Superata, P.J.Carroll, Phys.Rev. B3, 85 (1971); W.D.Gregory, Phys.Rev.Lett. 20, 53 (1968); W.D.Gregory, The Science and Technology of Superconductivity (Plenum Press, New York), 1973, vol. I, p. 211
/20/ This experiment will be discussed in detail by W.D.Gregory et al., later in this conference, paper C-14, this volume, p. 265
/21/ The important parameters are the threshold energy (or onset temperature) and the lineshape of the transition.
/22/ See, e.g. D.A.Hays, Phys.Rev. B1, 3631 (1970). Often the results of one experiment are used to "predict" the behavior of a different thermodynamic parameter and this prediction is

compared to experimental data; the degree of agreement is, then, used as a measure of validity of the value of $<a^2>$.

/23/a S.G.O'Hara, G.J.Sellers, A.C.Anderson, Phys.Rev. <u>B10</u>, 2777 (1974)

b T.Mamyia, A.Oota, Y.Masuda, Solid State Comm. <u>15</u>, 1689 (1974)

/24/ See, e.g. the appropriate entries in Table IV for Al, In, and Pb.

/25/ This is true except in the negative sense that no changes in the thermal conductivity with decreasing purity indicates major material problems as has been demonstrated for Ga, Al, and In. See Ref. /12/ and N.V. Zavaritskii, Soviet Phys. JETP <u>10</u>, 1069 (1960); <u>12</u>, 831 (1961).

/26/ In Fig.7b the fact that both pure and impure samples have the same ratio of electron phonon scattering (heavy line) reinforces this conclusion.

/27/ W.A.Phillips, Proc.Roy.Soc. (London) <u>A309</u>, 259 (1969)

/28/ To confirm the validity of these corrections, Phillips calculated the thermal conductivity appropriate for his UA measurements; these calculations compare well with measured thermal conductivity data /27/.

/29/ J.M.Perz, E.R.Dobbs, Proc.Roy.Soc. (London) <u>297</u>, 408 (1967)

/30/ J.A.Rayne, C.K.Jones, Physical Acoustics VII (Academic Press, New York), 1970, p. 149

/31/ It is certainly curious that a close look at the results of many different experiments shows that frequently two gaps are inferred for all crystalline orientations (an average Δ and a minimum Δ) such that the values of these gaps vary little, if at all, as a function of k-space orientation.

/32/a E.M.Forgan, C.E.Gough, J.Phys.F <u>3</u>, 1596 (1973)

b D.P.Almond, J.A.Rayne, J.Low Temp.Phys. <u>23</u>, 7 (1976)

/33/ At this time it is still not clear if UA experiments for Nb are consistent with zero gap anisotropy; however, see Ref. /32b/.

/34/ E.G.Brickwedde, D.E.Binnie, R.W.Reed, Proc. LT13, vol.III (Plenum Press, New York), 1974, p. 745; J.E.Randorff, B.J.Marshall, Phys.Rev. B2, 100 (1970)

/35/ From another point of view, however, it might be argued that because of the individual material problems involved in any given experiment, it is only in the limit of no anisotropy that a definitive statement can be made concerning energy gap anisotropy.

/36/ M.L.A. MacVicar, R.M.Rose, J.Appl.Phys. 39, 1721 (1968)

/37/ M.L.A. MacVicar, Phys.Rev. B2, 97 (1970)

/38/ S.I.Ochiai, M.L.A. MacVicar, R.M.Rose, Phys.Rev. B4, 2988 (1971)

/39/ J.Bostock, K.Agyeman, M.H.Frommer, M.L.A. MacVicar, J.Appl.Phys. 44, 5567 (1973)

/40/ N.V.Zavaritskii, Soviet Phys. JETP 18, 1260 (1964); Soviet Phys. JETP 21, 557 (1965)

/41/ C.K.Campbell, D.G.Walmsley, Can.J.Phys. 45, 159 (1967)

/42/ G.L.Wells, J.E.Jackson, E.N.Mitchell, Phys.Rev. B1, 3636 (1970)

/43/ B.L.Blackford, J.Low Temp.Phys. 23, 43 (1976)

/44/ G.I.Lykken, A.L.Geiger, K.S.Dy, E.N.Mitchell, Phys.Rev. B4, 1523 (1971)

/45/ B.L.Blackford, Physica 55, 475 (1971); B.L.Blackford, R.H.March, Phys.Rev. 186, 397 (1969); D.E.Banks, B.L.Blackford, Can. J.Phys. 51, 2505 (1973)

/46/ G.I.Rochlin, Phys.Rev. 153, 513 (1967)

/47/ W.D.Gregory, L.S.Straus, R.F.Averill, J.C.Keister, C.Chapman, Proc. LT 13, ed. by K.D.Timmerhaus et al. (Plenum Press, New York), 1974, vol.3, p. 316; J.C.Keister, L.S.Straus, W.D.Gregory, J.Appl.Phys. 42, 642 (1971); W.D.Gregory, R.F.Averill, L.S.Straus, Phys.Rev.Lett. 27, 1503 (1971); W.D.Gregory, The Science and Technology of Superconductivity, ed. by W.D.Gregory, W.N.Mathews, Jr.,

E.A.Edelsack (Plenum Press, New York), 1973, vol.I, p. 211
/48/ K.Yoshihiro, W.Sasaki, J.Phys.Soc.Japan 28, 262 (1970)
/49/ J.Polick, Ph.D. Thesis, Ohio State University (Department of Physics), 1973, unpublished
/50/ W.D.Gregory, Private Communication
/51/ A.Leger, J.Klein, Phys.Rev. B3, 3968 (1971)
/52/ K.R.Milkove, J.Bostock, M.L.A. MacVicar, to be published, Solid State Comm.

C-13 TUNNELING JUNCTION PHENOMENA: AN ANSWER TO UNANSWERED QUESTIONS*

M.L.A.MacVicar, J.L.Bostock and K.R.Milkove

Department of Physics, Massachusetts Institute of Technology, Cambridge, Massachusetts 02139, U.S.A.

In tunneling experiments whose objective it is to investigate the existence of multiple superconducting energy gaps and/or anisotropy of the superconducting energy gap value(s), data is obtained from individual tunnel junctions and displayed in the form of I-V and dI/dV (or dV/dI) characteristics (see Fig.1). The jump in current at the bias $(\Delta_A + \Delta_B)$ is a key feature in the curves and is used by experimenters to determine the values of the energy gaps Δ_A and Δ_B of the junction electrodes. It is general practice to design the junction to have one electrode exhibit a single-valued, isotropic energy gap (e.g., a thin film) so that all multiple gap and/or anisotropy effects in the data may be ascribed to the other electrode /1-7/. Such effects are thought to manifest themselves in the junction characteristics in the form of multiple current steps (at $\Delta_A^1 + \Delta_B$, $\Delta_A^2 + \Delta_B$, $\Delta_A^3 + \Delta_B$, etc.) corresponding to each gap value, and in the variation in magnitude of Δ_A^n with crystal orientation of the tunneling direction. Since much of the anisotropy literature is based largely on the interpretation of tunneling data, clearly, considerable care must be exercised in working backwards

---

*This work is supported by the U.S. Advanced Research Project Agency, N00014-75-C-1084

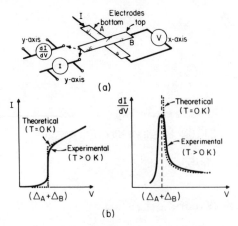

Fig.1: (a) Schematic of 4-point probe connections to a junction sandwich comprised of two metal electrodes separated by a very thin insulating layer (usually the natural oxide of the bottom electrode).

(b) Typical I-V and dI/dV-V characteristics expected for temperatures well below the superconducting transition temperatures of electrodes.

from tunneling characteristics to conclusions concerning fundamental energy gap properties.

Experimental I-V and dI/dV-V characteristics from Nb/Ox/In junctions are presented in Fig.2. (Previous research proves that Δ_{In} for such a junction is isotropic and single-valued /7/.) In Fig.2a two peaks are observed in the derivative near the jump in current of the I-V characteristic. Conventional analysis of this data leads us to the conclusion that Nb has two simultaneous gap values, Δ_{Nb}^1 and Δ_{Nb}^2, differing by 0.12 meV. Another set of I-V characteristics obtained from Nb/Ox/In junctions, perhaps at a different orientation, are presented in Fig.2b. These characteristics again show two current jumps, and, by usual data analysis procedures, lead to the conclusion that there are two gap values in niobium. One value is the same as in Fig.2a, i.e., $\Delta_{Nb}^{1'} = \Delta_{Nb}^{1}$; however, the second value, $\Delta_{Nb}^{2'}$, is quite different from the previous data. It would seem that niobium not only has two values of energy gap, but that one value is probably isotropic and the other, anisotropic.

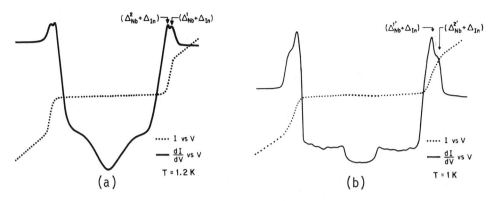

Fig.2a: I-V and dI/dV-V characteristics obtained from single-crystal-Nb/Ox/In-film-junctions. The derivative resolves two sum peak structures at biases $(\Delta_{Nb}^1+\Delta_{In})$ and $(\Delta_{Nb}^2+\Delta_{In})$, corresponding to apparent niobium gap values of $\Delta_{Nb}^1 = 1.54$ meV and $\Delta_{Nb}^2 = 1.42$ meV, respectively. ($\Delta_{In} = 0.55$ meV.)

Fig.2b: I-V and dI/dV-V characteristics obtained from single-crystal-Nb/Ox/In-film-junctions. The derivative resolves two sum peak structures at biases $\Delta_{Nb}^{1'} = 1.54$ meV and $\Delta_{Nb}^{2'} = 1.73$ meV, respectively. ($\Delta_{In} = 0.55$ meV.)

The appearance of the curves in Fig.2 are typical of curves appearing in the tunneling literature on multiple energy gaps and anisotropy. The question is whether the data really indicates (as the reported data on other materials purports to) that multiple energy gaps and/or anisotropy, in fact, exist in Nb. The answer is NO. The reason is that the data presented here is not what it appears to be. In fact, the data does not correspond to individual junctions, but to two Nb/Ox/In junctions connected in a special way.

Consider two separate, individual, ideal tunnel junctions each composed of material A as the lower electrode and material B as the upper electrode. For both junctions, the energy gaps Δ_A are identical and single-valued; similarly, the Δ_B's are identical and single-valued. We assume that each of these junctions is a perfect junction in that the current path between the electrodes A and B is a single path yielding characteristics as shown in Fig.1. Now, if these two junctions are connected in parallel, with a series resistance inserted in one parallel leg, and a systematic investigation of the electronic characteristics is performed, tunnel characteristics very similar to those just presented will be obtained.

There are two possible parallel configurations for the junctions according to where V is measured (see Fig.3). In case V_1, Fig.3, the I-V characteristics of the configuration must have the $(\Delta_A + \Delta_B)$ jump of the upper leg junction occuring <u>below</u> the $(\Delta_A + \Delta_B)$ jump of the lower leg junction by a bias equal to V_R which is the drop across the series resistor R. The V_2 configuration must have an I-V charac-

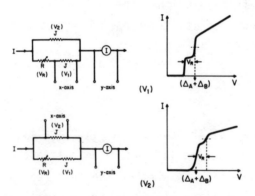

Fig.3: Possible configurations for two parallel tunneling junctions of identical geometries and gap values, arranged with a variable resistor, R, in series with one junction.
(V_1) Measuring voltage V_1 yields characteristics with structure observed just below the real sum peak bias, $(\Delta_A + \Delta_B)$.
(V_2) Measuring voltage V_2 yields characteristics with major structure observed just above the real sum peak bias, $(\Delta_A + \Delta_B)$.

teristic where the $(\Delta_A + \Delta_B)$ jump of the lower leg junction occurs <u>above</u> the $(\Delta_A + \Delta_B)$ jump of the upper leg junction by the bias V_R. For both V_1 and V_2 configurations, the <u>voltage difference</u> between the two peaks depends on the value of R relative to the junction resistance in series with it. The <u>relative amplitudes</u> of the two jumps, however, are further determined by the relationship of the total resistances of the upper and lower legs to one another for the temperature in question. (It is important to keep in mind that a superconducting tunnel junction is a non-linear, temperature dependent device.)

To obtain the data shown in Fig.2a, two Nb/Ox/In junctions were connected in the V_1 configuration. The resistance of the upper leg junction is 305 Ω; in the lower leg, the junction resistance is 32 Ω; the series resistor is 16 Ω. For this case, a resistance of 16 Ω is sufficient to split the derivative sum peak but is <u>not</u> sufficient to cause a clearly-developed step in I-V data. The effects on the I-V for a range of values of R is shown for the same configuration (V_1) in Fig.4. The upper leg junction is 16 Ω and the lower leg junction is 40 Ω. The data in Fig.2b is for two Nb/Ox/In junctions connected in configuration V_2. Here the upper leg junction resistance is 325 Ω; the lower leg junction resistance 200 Ω, and R = 33.5 Ω. A smaller value of R could also be chosen so that the derivative sum peak is not as split but is merely asymmetric and the I-V curve appears nearly ideal. For niobium, our data indicates that R$\sim$0.1Ω would be sufficient to produce a small but visible asymmetry in the sum peak of 100 Ω junctions.

There is nothing special in our choice of just two parallel legs in the discussion of the model presented here. The model is general and encompasses three, four, etc. .... legs with various resistors in different legs. The resulting I-V's and derivatives predicted by these more complex versions of the model have real-

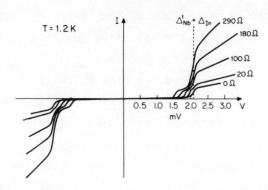

Fig.4: The I-V characteristic for two Nb/Ox/In junctions (17 Ω and 40 Ω) connected in configuration V_1 for various values of R in series with the 40 Ω junction.

data-counterparts in the reported and unreported literature, though not as many as for the two leg cases. (Researchers become increasingly mistrustful of data as the number of gaps it implies goes up.) It is our contention that investigators often observe I-V data such as that in Fig.2 while thinking that only a single junction (current path) is under study. We believe that this circumstance prevails in much of the anisotropic energy gap literature. Hence at the very least, all multiple gap-value data must be examined for its validity as an indicator of real gap values intrinsic to materials, and anisotropies associated with these multiple gap values cannot be assumed, a priori, to be valid. Also, data associated with anisotropic single gap value observations need to be examined in terms of deviations from ideal single current-path behavior.

References

/1/ N.V.Zavaritskii, Soviet Phys. JETP 16, 793 (1963); Soviet Phys. JETP 18, 1260 (1964); Soviet Phys. JETP 21, 557 (1965)
/2/ J.C.Keister, L.S.Straus, W.D.Gregory, J.Appl.Phys. 42, 642 (1971)
/3/ D.E.Banks, B.L.Blackford, Can.J.Phys. 51, 2505 (1973); B.L.Blackford, R.H.March, Phys.Rev. 186, 397 (1969); B.L.Blackford, Physica 55, 475 (1971)
/4/ C.K.Campbell, D.G.Walmsley, Can.J.Phys. 45, 159 (1967)
/5/ G.I.Lykken, A.L.Geiger, K.S.Dy, E.N.Mitchell, Phys.Rev. B4, 1523 (1971)
/6/ R.W.Cohen, B.Abeles, G.S.Weisbarth, Phys.Rev.Lett. 18, 336 (1967)
/7/ M.L.A.MacVicar, R.M.Rose, J.Appl.Phys. 39, 1721 (1968)

C-14 AN ANALYSIS OF EVIDENCE FOR SUPERCONDUCTING ENERGY GAP AND
PAIRING INTERACTION ANISOTROPY FOR TWO TYPES OF EXPERIMENTS*

W.D.Gregory, A.J.Grekas**, S.Horn+, L.Morelli

Department of Physics, Georgetown University

Washington, D.C. 20057, U.S.A.

1. Introduction

While there is a large body of evidence that the pairing interaction and energy gap in superconductors are anisotropic, there is also some inconsistency in many of the measurements of these quantities /1,2/. In this paper we would like to review and critique the results of two kinds of experiments, that we have performed over the last decade /3/, to measure possible anisotropy of the energy gap and pairing interaction in superconducting systems. These two techniques are (a) measurement of the shift in critical temperature due to removal of anisotropy by boundary scattering in single crystals or in well annealed polycrystalline foils, and (b) measurement of the energy gap via tunneling into single crystals of gallium.

---

*Work supported in part by NSF and ERDA.
**Present Address: Physics Department, City College of New York, New York, U.S.A.
+Present Address: Night Vision Laboratory, Fort Belvoir, Virginia, U.S.A.

2. Boundary Scattering Experiments

One of the most commonly employed techniques for estimating the pairing interaction anisotropy is the so-called "mean-free path effect". Performed in a variety of ways /4/, these experiments involve measuring one of several physical parameters /3a,3b,5/ (critical temperatures, heat capacity, critical field curve, spin-lattice relaxation time, thermal conductivity, etc.) that change when anisotropy is removed by limiting the electron mean-free path. The mean-free path has been limited in experiments by impurity addition, boundary scattering and introduction of defects in the material. Markowitz and Kadanoff /4/, and later Clem /5/, have analyzed this type of experiment extensively, using a form of pairing interaction of a superconductor that is separable, with anisotropy factors as a function of crystalline direction $a(\underline{\Omega})$.

By far the largest body of evidence for anisotropy in superconductors comes from the mean-free path effects obtained from the shift in <u>critical temperature</u> with addition of impurities. These experiments often lack consistency and have a philosophical difficulty, in that the addition of impurities produces chemical effects, strains, damage and other less-than-ideal conditions that can shift the critical temperature more than the reduction of T_c due to anisotropy averaging. The situation is even less clear with the introduction of a reduced mean-free path via damage and cold working.

A much cleaner experiment to examine the mean-free path effect is to limit the mean-free path via scattering from the boundary of the material. This has been performed with small <u>single crystal</u> materials /3a,c,d/, allowing one to avoid alloying and to use samples that are as free from strain and damage as possible. Even when pounded foils are used, because no alloying is involved, as much annealing as is desired may be used to remove residual shifts of critical temperature due to strain. In addition, the parameter λ^i in the Markowitz-Kadanoff formulation can be estimated much more readily for the boundary scattering technique.

Estimates of λ^i in the case of impurities is an impossible task and, therefore, it was used exclusively as a fitting parameter in comparing the Markowitz-Kadanoff formulation to impurity addition data (values ranging from 0.4 to 1.0 were obtained). However, the integrals <u>can</u> be calculated for the boundary scattering situation /3c/ with an uncertainty only in the degree to which the scattering from the surface must be considered diffuse or specular.

In addition to choosing the boundary technique itself as a cleaner method for estimating the mfp effect, we also chose to measure the shift in critical temperature by the mutual inductance method, which measures the change in the mutual inductance of a pair of coils, containing the sample, due to the expulsion of flux by the Meissner effect at the critical temperature. It is possible with this technique to use magnetic fields in the primary coil that do not shift T_c within the precision of the measurement (10^{-4} K). Also, this will be essentially a volume technique, if low frequencies are used, such that the depth of penetration of the a.c. field is of the order of the sample thickness, when the sample is in the normal state. Fortunately, this is possible in these experiments, since the samples employed were quite thin (5 - 500 μm).

The apparatus and experimental techniques have been described previously /3a/. All temperature measurements were made relative to the T_c of a bulk specimen and the apparatus was contained in a series of mumetal and conetic shields, or in a Helmholtz coil pair, to reduce the magnitude of the earth's field. (With shielding, the earth field was reduced to approximately 3 milligauss, and with the Helmholtz pair, to approximately 10 milligauss, in typical experiments.) A number of control experiments were performed in which the specimens were heated back and forth through T_c at various rates (without noticeable hysteresis) and in which the specimens were tilted at various angles in the residual magnetic field. These experiments eliminated the possibility that thermal gradients or residual magnetic field effects affected the data.

The samples were of two kinds - single crystals which were grown in a specially-designed mold /3e/ and seeded to single crystal growth (gallium and indium) and pounded and annealed polycrystal foils /8/ (indium, zinc, aluminium, tin and lead). Sample preparation and the weight averaging method for measuring specimen thickness have been described previously /3a,e,f/.

For polycrystalline specimens, a number of auxiliary experiments were performed. The samples were annealed and remeasured many times, increasing the annealing temperature and the annealing time until the critical temperature of the specimen had stopped changing. (An example of this is shown for typical data in Fig.1.) Secondly, because of the possibility of contamination during the pounding to a thin foil, the polycrystalline specimens were subjected to additional spectroscopic analyses and Auger surface spectra were acquired. The results of these analyses indicated that metallic impurities were present in amounts less than 0.01%, too little to cause the observed shifts in T_c.

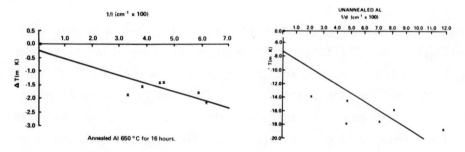

Fig.1: Comparison of the shift in T_c with specimen thickness for Al, (a) annealed and (b) unannealed

The single crystal specimens were of higher metallurgical quality. We used these to estimate the effects of (1) <u>residual</u> strain after annealing and (2) strain effects caused by the epoxy glue used to mount the specimens. We found that single crystal Ga and In plates could be bent back and forth, annealed and remeasured, with the mutual inductance traces and estimated T_c values repeating to within the 10^{-4} K precision of the measurement. The epoxy was found to produce a smearing of the transition on the normal state side, but the predominant jump in mutual inductance was centered about the same temperature value to within $\pm 10^{-4}$ K, as the proportion of the sample covered by epoxy was increased. Details of these experiments are discussed elsewhere /3a/.

We also attempted to measure the resistance ratio of a number of the specimens. However, we discovered that, with specimens of this size, the difficulty of leads placement in obtaining a true resistance measurement unaffected by current streaming problems /9/ was such that a more accurate and reliable estimate of the mean-free path could be obtained from the chemical analysis and the over-all behavior of the data. Therefore, a check was made that we obtained a true boundary scattering limit on the mfp by comparing single crystal and polycrystal indium data to the correct transport expression for the boundary limited mfp. This is, in fact, another virtue of the boundary scattering method, since the connection between the mean-free path and the specimen thickness is logarithmic. As given by Fuchs /10/, corrected for the free path averaging appropriate to anisotropy removal by Gregory /3a,c/, this relation is

$$\ell = \frac{1}{2} d \ln(\ell_0/d) \qquad \begin{aligned}\ell_0 &= \text{bulk m.f.p.} \\ d &= \text{sample thickness}\end{aligned} \qquad (1)$$

One can see that the shifts of critical temperature observed are a truly mean-free path phenomenon, as illustrated in Fig.2. A summary of all our boundary scattering results are given in Fig.3 and Table 1.

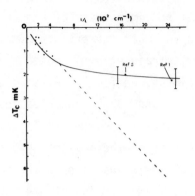

Fig.2: Shift in T_c with inverse boundary limited mean-free path (solid line) and inverse specimens thickness (dashed line) compared to data from several sources. Refs./1/ and /5/ refer to notation of ref./3d/

3. Bulk Tunneling

One of the potentially most useful techniques to probe energy gap anisotropy would seem to be tunneling into single crystals, a technique that might yield resolution of the energy gap structure over small cones of angles in k space. Following our earlier report of successful tunneling into single crystals of gallium, indium and aluminium /3b,g,h,1/, we decided to design and conduct experiments to see if some of the difficulties with energy gap measurements from bulk tunneling could be eliminated. Our goal in this study was to develop a technique whereby the precision, with which the energy gap could be measured, would be 1% or better. But, more important, we required a technique that would allow internal statistical checks on the quality of the I-V curves.

The choice of the material to tunnel into was based on a series of considerations. Bulk tunneling had been successfully carried out in tin /11/, tantalum /12/, niobium /13/, lead /14/, aluminium /15/, indium /16/, and gallium /17,3/.

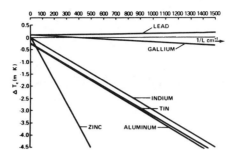

Fig.3: Summary of the data on shift of T_c with boundary limited meanfree path. Errors tabulated in Table I.

Table I: Anisotropy and Scattering Effects. Data are expressed as least-square fits to $\Delta T_c = A + B/\ell$

Metal	$<a^2>$[a]	$<a^2>_\lambda^i$[b]	λ^i	A(mK)	B x 10^3 (mK-cm)	Correlation[c] Coefficient
Al	0.016	0.011	0.7	-0.24 ± 0.29	-3.02 ± 0.64	-0.89
In	0.027	0.018	0.7		-2.94 ± 0.15	
Pb	0.00	<0.001		0.09 ± 0.14	0.069 ± 0.18	-0.15
Sn	0.016	0.0138	0.9	-0.22 ± 0.29	-2.22 ± 0.47	-0.89
Zn	0.031	0.029	0.9		-9	d
Ga	0.008	0.0053	0.7	0.1 ± 0.04	-0.59 ± 0.57	

[a] critical field data, ref. /24/
[b] boundary scattering
[c] related to probability of a true linear fit (see text)
[d] the data on Zn are insufficient to quote an estimated error

The overriding consideration was based on the fact that the superconductors for which a theory about the gap anisotropy had been put forward /4,5/ are weak-coupling. Just as important, the material must be 1) very pure metallurgically, since impurities in exceedingly small amounts are known to affect drastically the tunneling I-V curves for some materials, 2) be nonreactive, 3) be easily grown in any desired crystal orientation, 4) be easily annealed close to its melting point, as stresses are known to affect interpretation of the data.

Some other ancillary considerations were: a) as much prior data as possible available in the literature, b) expertise within our laboratory with the material to be chosen.

A metal like niobium was not considered as it fails 1), 2), 3), and 4). It is not clear that lead is anisotropic; aluminium is very sensitive to stresses /3f/; hence tin, indium and gallium were then the candidates left. Gallium has all the properties required, some of which are perhaps worth mentioning all in one place: It is commercially available in chemical purity exceeding 99.9999%, its residual resistance ratio is between $10^6 - 5 \times 10^6$, it melts at 29.8 °C, it can be easily injected in a mold, super cooled and grown as a single crystal in any desired orientation using a seed crystal, it zone refines while growing, thus rejecting impurities, it anneals as it grows and keeps on annealing until cooled in the cryostat.

All of these desirable features, plus the data available in the literature and our own expertise in handling it, made gallium the obvious choice. In designing the experimental arrangement, we sought to eliminate difficulties that had become apparent from earlier studies:

<u>AC and RF effects.</u> In recent investigations we have found that very small AC or RF pickup can shift the voltages measured in an apparently DC measurement of I-V characteristics enough to make it imperative to reduce AC ground loops and RF pickup below voltages required for the precision of the measurement desired. Because of these effects, we also felt it best to stay away from AC derivative techniques.

S-S' Tunneling vs N-S Tunneling. For S-S' tunneling, a widespread method to evaluate the energy gap is based on the presence of the sum and difference of the gaps. Since the difference cusp is only observable at temperatures near T_c, S-S' junctions have the disadvantage that one must extrapolate to zero temperature from somewhat higher experimental temperatures, with attendant errors. In contrast, N-S junctions have, on the I-V curves, better defined features as $T \to 0$.

Sample preparation. We have discovered that it is possible, using gallium, to make N-S tunnel junctions of a higher "quality" than S-S' tunnel junctions in terms of yield, sample reproducibility and I-V curves. Thus, we proceeded with the study using a normal metal (silver) counter electrode on gallium, with the sample grown in a temperature, humidity and dust controlled room.

Data Acquisition. Before running a batch of samples, the procedure used to eliminate AC ground loops was to examine the characteristics of a dummy sample placed in the sample position with a HR 8 lock-in amplifier phased to the 60-cycle line current. In addition, the experiment was performed in a shielded room, which showed by test to be good to approximately 100 db shielding at 146 MHz. To avoid the temperature extrapolation problem, measurements were made in a dilution refrigerator with T/T_c of the order of 0.10. The data were taken with a simple DC bridge and no AC electronics were used in the circuit.

Criterion for determining the energy gap. We continued to use the method of Shapiro et al. /18/ in determining the energy gap by fitting a BCS based tunnel integral to the tunneling curves. The purpose of this was to use as large a portion of the I-V curve to define the energy gap as possible, rather than a few selected points. The data were digitized and fit to the theoretical I-V curves /3j/ with standard statistical techniques /19/, the energy gap being the fitting parameter. Fig.4a shows a typical fit between experimental and theoretical I-V curves. Fig.4b shows the derivative curve computed numerically from the best theoretical fit to the I-V curve for the sample shown in Fig.4a and the position of the best fit energy gap is also shown for reference.

Seventeen N-S silver (film)-gallium (crystal) tunnel junctions were measured at 12 different orientations in the gallium AB plane.

Data Analysis. In the analysis, data were accepted only if the sample passed the following tests. First of all, the procedure for eliminating ground loops described above had to be passed successfully. In addition, the same I-V curve had to be repeatable, within

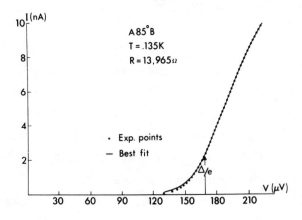

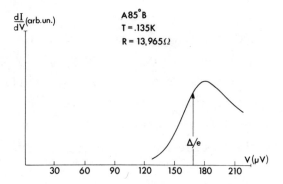

Fig.4: Tunneling data on Ga using Ag counter-electrode (N-S tunneling). (a) I-V curve (b) Best fit and data derivative curve

the standard deviation of the fit, for both polarities of the measuring voltage and after several temperature excursions from room temperature to cryogenic temperatures. We found that, as an internal criterion for sample quality, these criteria were satisfied when we obtained a precision of the gap, as a fitting parameter, of 0.03 - 0.05%. (Note carefully that this was not the repeatability from junction to junction, but for one junction measured at different times.) Of 17 samples, 15 fit these criteria.

Figure 5 shows the results for all 15 samples, with the energy gap plotted as a function of the angle perpendicular to the sample surface, as measured from the gallium A axis. Samples that were measured on more than one occasion are marked by the letter A. These data repeated within the precision of the fit ($3 - 5 \times 10^{-2}$%). A few samples, made from different batches, were measured at the same angles and they are marked by the letter B. This repeatability, from sample to sample, was of the order of 0.3%.

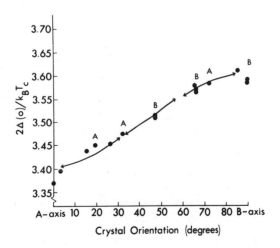

Fig.5: Summary of the zero temperature gap vs. crystal orientation in the Ga AB plane obtained from N-S data

4. Discussion and Conclusions

4.1 Boundary Scattering Techniques

The results from boundary scattering indicate an anisotropy of the pairing interaction outside the probable errors in the measurement for indium, tin and aluminium. The correlation coefficient (a measure for ΔT_c and ℓ to vary together) is very high for aluminium and tin, while in the case of lead it has a low value, consistent with no gap anisotropy present in the measurement. For zinc, the paucity of data precludes an estimate of the errors, but the preliminary indication is that the boundary scattering shift in T_c is substantial. The value of the mean square anisotropy for In, Sn and Al is, within the statistical error, close to that estimated for a number of impurity addition experiments. Gallium exhibits little anisotropy; essentially zero is observed within the uncertainty of the measurement. However, the anisotropy predicted for gallium from the bulk tunneling experiments discussed in this paper is sufficiently small that there is no inconsistency between the boundary scattering and bulk tunneling experiments. For lead there is a possibility that, because of the very small coherence length, the shift in critical temperature due to boundary scattering would be very small in the range of specimen thicknesses used, since the shift in critical temperature is proportional to an inverse mean-free path normalized by the coherence length.

The effect of stresses can be serious for some of these materials (e.g. Al) and not for others (e.g. Ga); the variety of control experiments that we conducted shows that, after prolonged annealing close to the melting point, the stresses can be removed so as not to obscure the behavior of the boundary scattering data, even for extreme sample mistreatment as the bending back and forth of the Ga plates mentioned in section 2.

Hence, because of the cleanliness of the boundary scattering experiments and because of the control experiments performed, we believe that boundary scattering results offer affirmative evidence for energy gap and pairing interaction anisotropy in superconductors that does not have some of the difficulties of other techniques, such as the bulk tunneling method to be discussed below. On the other hand, the precision of the measurements is such that materials with small anisotropy (equivalent to the energy gap anisotropy of less than 10%) would appear to have negligible anisotropy within the precision of the measurement. We believe that this reconfirmation of the average anisotropy in those materials that did exhibit large shifts in critical temperature is sufficiently strong to warrant further theoretical investigations to estimate the average anisotropy for specific materials.

4.2 Bulk Tunneling Technique

While it would seem advisable to try to extend the boundary scattering method and try to obtain data with greater precision, this seems unlikely to bear fruit, for two reasons: (1) the experiments described were performed with about as much precision as the existing technology would allow and, (2) in any event, boundary scattering yields only a value of the _average_ anisotropy. We would see a role for boundary scattering as a convenient check of the overall average anisotropy, as obtained from a more precise measurement, such as bulk tunneling, and would urge some effort to be made to answer some of the concerns regarding bulk tunneling, particularly those we have heard at this meeting.

The present status of our experiments in bulk tunneling indicates that N-S tunneling at low temperature, with digitized DC electronics in a shielded environment, may enable one to measure apparent energy gaps with a repeatability substantially better than 1%. We note from Fig.5 that the overall anisotropy is of the order of 7%, in agreement with previous measurements /3b/. In the present case, the anisotropy appears to be centered about an energy gap value somewhat lower than the extrapolated zero gap obtained from the S-S' measurements. While it is not certain at this time, we believe this may be simply due to some systematic errors in extrapolating the S-S' data to lower temperatures. (We also note that a deviation from BCS energy gap temperature dependence is expected for a truly anisotropic pairing interaction.)

The solid line in Fig.5 is a theoretical expression for the anisotropic energy gap for gallium in this plane obtained by Clinton and Cirillo /20/ and discussed in the following paper. We note that there is good agreement with the theory, within the repeatability of these data, for those regions of angles where the theory is applicable. One will note that there are certain regions of angles for which a solid line does not exist in Figure 5. In those areas, the k vector perpendicular to the sample surface crosses a Brillouin zone in the extended zone model. At such points, the theory in the following paper is not applicable, but instead we expect some discontinuity in the energy gap /3b,21/. Such discontinuities were observed at 32, 61, 70, and 80 degrees in two of the previous measurements /3b, 17/. The data of this work are not sufficiently detailed in those regions to reconfirm those observations and we have further work in progress to check closely the gap at these angles. Also, although these data seem quite good, we would caution the reader that considerable more data are necessary in order to confirm the repeatability of 0.3% quoted above.

Further, the following areas of concern regarding bulk tunneling must be addressed: (1) The possible counter electrode phenomena, discussed in the previous paper by Bostock and MacVicar /2/, should be determined for N-S tunneling using the digital technique. In this regard, we have begun a program to compare our results to the more standard derivative techniques to see if any of the effects described by Bostock and MacVicar are present. (2) There is a problem of eliminating possible systematic variations of the barrier structure, as proposed by Dowman et al. /22/. (3) Finally, we quote some recent work of our group in studying the properties of inhomogeneous superconducting films /23/. In this work we have found that the electrodynamics of inhomogeneous films, particularly those that are partially superconducting, can be markedly different from those of an assumed homogeneous film. In particular, one can observe variations of the critical temperature (and possibly the energy gap) across the films which would tend to smear the data and produce erroneous measurements of the gap parameter.

While we do not have any immediate answers to these problems, we would point out that the N-S results quoted in this work may take a first step in answering these questions. In general, the good agreement between the data reported here, with silver as a counter electrode, and the previous data /31/ with lead as a counter electrode would seem to reduce somewhat the concern over serious counter electrode effects. Also, the overall higher quality of the I-V curves for N-S junctions may be an indication that inhomogeneity phenomena in normal metal films are less serious than in superconducting counter electrodes, indicating normal-superconductor tunneling as a method of choice. This result would be consistent with the known variations of the electrodynamic properties of inhomogeneous films from the effective medium theory /23/. As a closing remark, we would like to stress that correct choice of the superconductor and careful experimental techniques are of paramount importance in bulk tunneling, lest one wishes to incur in results which are just artifacts of samples not good enough for bulk tunneling. Our conclusion is that problems, as the ones pointed out by ref. /2/, can and must be avoided.

The authors are indebted to a large group of co-workers and graduate students that passed through the low temperature group at Georgetown during the last decade for numerous discussions and an all-important helping hand in the laboratory. The members of this group are too numerous to mention individually.

References

/1/ A comprehensive review of the evidence for anisotropy of the superconducting energy gap is given by A.G.Sheplev, Soviet Phys. Usp. 11, 690 (1969). Also see Ref. /2,3a,3f/

/2/ J.L.Bostock, M.L.A.MacVicar, paper R-6, this volume, p.213

/3/ Over the last decade a number of experimental papers on anisotropy have been published by the Low Temperature Group at Georgetown University. Theses, when unpublished, are available from Dissertation Copies, P.O.B. 1764, Ann Arbor, Michigan 48106

   a W.D.Gregory, Ph.D.Thesis, MIT 1966 (unpublished); W.D.Gregory, T.P.Sheahen, J.F.Cochran, Phys.Rev. 150, 315 (1966)

   b W.D.Gregory, L.S.Straus, R.F.Averill, J.C.Keister, C.Chapman Low Temp. Phys. LT 13, vol.3, p. 316 (Ed. Timmerhaus and Hammel, Plenum Publ.Corp.) 1972

   c W.D.Gregory, Phys.Rev.Lett. 20, 53 (1968)

   d W.D.Gregory, M.A.Superata, P.J.Carroll, Phys.Rev. B3, 85 (1971)

   e W.D.Gregory, M.A.Superata, J.Cryst.Growth 7, 5 (1970)

   f S.B.Horn, Ph.D. Thesis, Georgetown University 1974 (unpublished)

   g J.C.Keister, L.S.Straus, W.D.Gregory, J.Appl.Phys. 42, 642 (1971)

   h R.F.Averill, L.S.Straus, W.D.Gregory, Appl.Phys.Lett. 20, 55 (1972) i W.D.Gregory, R.F.Averill, L.S.Straus, Phys.Rev.Lett. $\underline{27}$, 1503 (1971)

  j A.Grekas, Ph.D. Thesis, Georgetown University 1976 (unpublished)

/4/ D.Markowitz, L.P.Kadanoff, Phys.Rev. $\underline{131}$, 563 (1963)

/5/ J.Clem, Annals of Phys. $\underline{40}$, 268 (1966)

/6/ E.A.Lynton, D.McLachlan, Phys.Rev. $\underline{126}$, 40 (1962)

/7/ For example, see R.Oder, Ph.D. Thesis, MIT 1964 (unpublished)

/8/ S.B.Horn, W.D.Gregory, to be published (see also Ref. /3f/)

/9/ D.K.C. MacDonald, Phil.Mag. $\underline{2}$, 97 (1957)

/10/ K.Fuchs, Proc.Cambridge Phil.Soc. $\underline{34}$, 100 (1938)

/11/ N.V.Zavaritskii, Soviet Phys. JETP $\underline{18}$, 1260 (1964)

/12/ W.D.Sheril, H.E.Edwards, Phys.Rev.Lett. $\underline{6}$, 460 (1961). See also I.Dietrich, Z.Naturforsch. $\underline{17a}$, 94 (1962) and I.Dietrich, Proc. VIII Internat. Conf. Low Temp. Phys., London (1963), p. 173

/13/ M.L.A.MacVicar, R.M.Rose, J.Appl.Phys. $\underline{39}$, 1721 (1968). See also M.L.A.MacVicar, R.M.Rose, Phys.Lett. $\underline{25A}$, 681 (1967)

/14/ B.L.Blackford, R.H.March, Phys.Rev. $\underline{186}$, 387 (1969)

/15/ W.D.Gregory, L.S.Straus, R.F.Averill, J.C.Keister, C.Chapman, Low Temp. Phys. LT 13, vol.3, p.316 (Ed. Timmerhaus and Hammel, Plenum Publ.Corp.) 1972. See also B.C.Blackford, Dalhousie University, to be published (1975)

/16/ R.F.Averill, L.S.Straus, W.D.Gregory, Appl.Phys.Lett. $\underline{20}$, 55 (1972). See also Ref. /3b/

/17/ See Ref. /3b,3g,3i/. See also K.Yoshihiro, W.Sasaki, J.Phys. Soc.Japan $\underline{24}$, 426 (1968), K.Yoshihiro, W.Sasaki, J.Phys. Soc.Japan $\underline{28}$, 262 (1970), J.J.Polick, Ph.D. Thesis, Ohio State University 1973 (unpublished)

/18/ S.Shapiro, P.M.Smith, J.Nicol, J.L.Miles, P.F.Strong, IBM J. Res.Develop. $\underline{6}$, 34 (1962)

/19/ P.R.Benington, "Data Reduction and Error Analysis for the Physical Sciences", McGraw-Hill, New York 1969

/20/ N.C.Cirillo,Jr., W.L.Clinton, paper C-15, this volume, p.283
/21/ I.Wu, W.L.Clinton, to be published
/22/ J.E.Dowman, M.L.A.MacVicar, R.J.Waldram, Phys.Rev. 186, 452 (1969)
/23/ R.Janik, L.Morelli, N.C.Cirillo,Jr., J.N.Lechevet, W.D.Gregory, W.L.Goodman, IEEE Transactions on Magnetics, Mag. 11, 687 (1975). See also W.D.Gregory, L.Morelli, Bull.Am.Phys.Soc. 21, 403 (1976)
/24/ D.U.Gubser, Phys.Rev. B6, 827 (1972).

C-15 A NEARLY FREE ELECTRON MODEL OF THE BCS GAP EQUATION:
ENERGY GAP ANISOTROPY IN GALLIUM*

N.C.Cirillo,Jr. and W.L.Clinton
Department of Phyics, Georgetown University
Washington, D.C. 20057, U.S.A.

1. Introduction

Experimental manifestations of superconductor energy gap anisotropy have been observed in data for low temperature specific heat, the critical field, nuclear spin-lattice relaxation time, tunneling, surface resistance, and longitudinal ultrasonic attenuation /1/. Motivated largely by the tunneling experiments reported in the preceding paper /2/, a model for gap anisotropy in pure, single crystal, weak-coupling superconductors is presented.

The theoretical basis for energy gap anisotropy has its origins in the BCS gap equation /3/

$$\Delta_{\underline{k}} = -\sum_{\underline{k}'} V_{\underline{k},\underline{k}'} \frac{\Delta_{\underline{k}'}}{2E_{\underline{k}'}} \tanh(\tfrac{1}{2}\beta E_{\underline{k}'}) \qquad (1)$$

where $\beta = (k_B T)^{-1}$, $E_{\underline{k}} = (\xi_{\underline{k}}^2 + \Delta_{\underline{k}}^2)^{1/2}$, and $\xi_{\underline{k}}$ is the normal state energy of a Bloch electron relative to the Fermi surface. The entire $\underline{k}$ dependence in the gap parameter may be found in the pairing interac-

---

*Work supported in part by the National Science Foundation.

tion matrix element $V_{\underline{k},\underline{k}'}$ which is given by

$$V_{\underline{k},\underline{k}'} = \int d^3 r_1 \, d^3 r_2 \, \phi_{\underline{k}}^*(\underline{r}_1) \phi_{-\underline{k}}^*(\underline{r}_2) V_{12} \phi_{\underline{k}'}(\underline{r}_1) \phi_{-\underline{k}'}(\underline{r}_2). \tag{2}$$

$\phi_{\underline{k}}(\underline{r})$ is a Bloch wavefunction and V_{12} is the effective electron-electron potential appropriate for a superconductor.

One of the more useful methods of analysis of superconductor anisotropy is due to Markowitz and Kadanoff /4/ (MK). In order to study the effects of anisotropy on the critical temperature, MK assumed that the pairing interaction $V_{\underline{k},\underline{k}'}$ had the form

$$-V_{\underline{k},\underline{k}'} = (1 + a_{\underline{k}}) V (1 + a_{\underline{k}'}) \tag{3}$$

where V is the BCS constant attractive interaction. The anisotropy function $a_{\underline{k}}$ was assumed to depend only on the orientation of the vector $\underline{k}$ and it was chosen so that it averaged to zero over the Fermi surface, i.e., $\langle a \rangle \equiv \int (d\Omega_{\underline{k}}/4\pi) a_{\underline{k}} = 0$ for $|\underline{k}| = k_F$. It can then easily be inferred from Eq.(1) that

$$\Delta_{\underline{k}} = \langle \Delta \rangle (1 + a_{\underline{k}}). \tag{4}$$

Using the MK factorable form for the pairing matrix element $V_{\underline{k},\underline{k}'}$, Clem /5/ was able to theoretically examine the role of the angular dependence of $\Delta_{\underline{k}}$ in a variety of experimental quantities. Many of the interesting superconductor properties were seen to depend on either the anisotropy function $a_{\underline{k}}$ or on its mean-square average $\langle a^2 \rangle$. Experimental estimates of $\overline{a}_{\underline{k}}$ indicate that it could vary by ± 0.2 as a function of the direction of $\underline{k}$ in a weak-coupling superconductor; typical values of $\langle a^2 \rangle$ are in the range 0.02. These values are corroborated in much of the data summarized by Shepelev /1/.

A more detailed discussion of energy gap anisotropy was given by Bennett /6/ and Leavens and Carbotte /7/ in the context of the Eliashberg equations. These authors applied their results to lead and aluminium. Although offering the promise of more accuracy their method lacks the attractive simplicity of the MK method. On the other hand, although much experimental data has been analyzed in terms of either a_k or $\langle a^2 \rangle$, the MK ansatz, Eq.(3), has never been related to a microscopic theory. We will show in the present work that the simplicity of Eq.(3) is retained and a microscopic expression for a_k derived when one uses nearly free electron (NFE) Bloch functions along with a reasonable choice for $V_{12} \equiv V(\underline{r}_1,\underline{r}_2)$ in Eq.(2).

2. A Nearly Free Electron Model

In his original work in superconductivity, in which he introduced the factorization of the two-body propagator, Gor'kov /8/ used a simplified two-body potential

$$V(\underline{r}_1,\underline{r}_2) = -V\delta(\underline{r}_1 - \underline{r}_2)\Omega \tag{5}$$

where V is a positive constant and Ω is the volume of the system. Equation (5) represents an attractive effective potential which, because it is translationally invariant, cannot exhibit the effects expected in a real crystal. For example, one might expect the strength of the two-body potential to depend on whether it is in an ionic core or interstitial region. We thus propose the modification

$$V(\underline{r}_1,\underline{r}_2) = -V(\underline{r}_1)\delta(\underline{r}_1 - \underline{r}_2)\Omega \tag{6}$$

where $V(\underline{r}_1)$ is a real one-body potential function which exhibits the crystal symmetry. By integrating Eq.(6) over all space, we see that $V(\underline{r})$ may be interpreted as a spatial average of the full two-body potential. Using Eq.(6) the pairing interaction for the case where there are no magnetic fields becomes

$$-V_{\underline{k},\underline{k}'} = \Omega \int d^3r \, |\phi_{\underline{k}}(\underline{r})|^2 V(\underline{r}) |\phi_{\underline{k}'}(\underline{r})|^2. \tag{7}$$

It is now evident that there are two quantities effecting the anisotropy in the pairing interaction: the average two-body potential $V(\underline{r})$ and the Bloch function $\phi_{\underline{k}}(\underline{r})$.

We next calculate $V_{\underline{k},\underline{k}'}$ using NFE perturbation Bloch wavefunctions. In order to do this we must distinguish two regions of interest in $\underline{k}$-space: the first occupies nearly all of the area of the Fermi surface and comprises the part not near a Brillouin zone boundary. The second region includes the remaining portions of the Fermi surface which are near zone boundaries. While we may restrict our study of $\Delta_{\underline{k}}$ to either of these two regions, it is important to note that the $\underline{k}^T$ summation in Eq.(1) includes both. We will assume, however, that in doing the summation over the entire Fermi surface the error made in neglecting the second region is small.

2.1 NFE Model Away from Brillouin Zone Boundaries

In this case the Bloch wavefunctions to first order in the normal state pseudopotential $U(\underline{g})$ are

$$\phi_{\underline{k}}(\underline{r}) = \frac{e^{i\underline{k}\cdot\underline{r}}}{\Omega^{1/2}} + \sum_{\underline{g}}{}' \frac{U(\underline{g})}{\varepsilon_{\underline{k}} - \varepsilon_{\underline{k}+\underline{g}}} \frac{e^{i(\underline{k}+\underline{g})\cdot\underline{r}}}{\Omega^{1/2}} \qquad (8)$$

where the summation over reciprocal lattice vectors $\underline{g}$ does not include zero and $\varepsilon_{\underline{k}}$ is the unperturbed energy of an electron relative to the Fermi surface. Using Eqs.(7) and (8), we may easily calculate the matrix element $V_{\underline{k},\underline{k}'}$. We find

$$-V_{\underline{k},\underline{k}'} = V(0) + \sum_{\underline{g}}{}' U(\underline{g}) V(\underline{g}) \{\chi_{\underline{k},\underline{g}} + \chi_{\underline{k}',\underline{g}}\} \qquad (9)$$

where

$$\chi_{\underline{k},\underline{g}} = (\varepsilon_{\underline{k}} - \varepsilon_{\underline{k}-\underline{g}})^{-1} + (\varepsilon_{\underline{k}} - \varepsilon_{\underline{k}+\underline{g}})^{-1}, \qquad (10)$$

$$V(\underline{g}) = \int \frac{d^3r}{\Omega} V(\underline{r}) e^{-i\underline{g}\cdot\underline{r}}. \qquad (11)$$

We note that $V_{\underline{k},\underline{k}'}$ has the separable form $-V_{\underline{k},\underline{k}'} = V_{\underline{k}} + V_{\underline{k}'}$ where

$$V_{\underline{k}} = \frac{1}{2} V(0) + \sum_{\underline{g}}' U(\underline{g}) V(\underline{g}) \chi_{\underline{k},\underline{g}}. \qquad (12)$$

We can facilitate our comparison to Markowitz-Kadanoff by rewriting $V_{\underline{k},\underline{k}'}$ as

$$-V_{\underline{k},\underline{k}'} = 2<V> \{ 1 + \frac{V_{\underline{k}} - <V>}{2<V>} + \frac{V_{\underline{k}'} - <V>}{2<V>} \}. \qquad (13)$$

The comparison is completed by noting that Eq.(3) has the approximate form $-V_{\underline{k},\underline{k}'} \simeq V(1 + a_{\underline{k}} + a_{\underline{k}'})$. We can now define

$$a_{\underline{k}} \equiv \frac{V_{\underline{k}} - <V>}{2<V>} = \frac{V_{\underline{k}} - <V>}{V(0)}. \qquad (14)$$

Eq.(14) represents therefore the correct Markowitz-Kadanoff $a_{\underline{k}}$ through first order in the normal state pseudopotential $U(\underline{g})$. Since only the quantity $\chi_{\underline{k},\underline{g}}$ depends on $\underline{k}$, Eq.(14) takes the form

$$a_{\underline{k}} = \sum_{\underline{g}}' U(\underline{g}) \frac{V(\underline{g})}{V(0)} \{ \chi_{\underline{k},\underline{g}} - <\chi_{\underline{g}}> \}. \qquad (15)$$

Note that $<a> = 0$. The Fermi surface average of $\chi_{\underline{k},\underline{g}}$ is easy to perform and we obtain

$$<\chi_{\underline{g}}> = \frac{1}{gk_F} \ln \left| \frac{1-\alpha}{1+\alpha} \right|, \qquad \alpha \equiv g/2k_F. \qquad (16)$$

All of the physical quantities in Eq.(15) are known with the exception of the parameter $V(\underline{g})/V(0)$. We will return to a more detailed analysis of the anisotropy function $a_{\underline{k}}$ in section 3.

2.2 NFE Model on a Zone Boundary

We now wish to extend the discussion of the previous section to include regions in $\underline{k}$-space near Brillouin zone boundaries. In such regions, the Bloch wavefunctions may be approximated by

$$\phi_{\underline{k}}(\underline{r}) = C_{\underline{k}} \frac{e^{i\underline{k}\cdot\underline{r}}}{\Omega^{1/2}} + \sum_{\underline{g}}{}' C_{\underline{k},\underline{g}} \frac{e^{i(\underline{k}+\underline{g})\cdot\underline{r}}}{\Omega^{1/2}} . \tag{17}$$

The sum over $\underline{g}$ runs over all nonzero reciprocal lattice vectors $\underline{g}$ such that $|\underline{k}| = |\underline{k}+\underline{g}|$.

For simplicity we will consider the case where there is strong mixing only between the two states $e^{i\underline{k}\cdot\underline{r}}$ and $e^{i(\underline{k}-\underline{G})\cdot\underline{r}}$. Thus, we are restricting ourselves to regions in $\underline{k}$-space where the two plane wave model is valid. Denoting any $\underline{k}$ which lies directly on the Brillouin zone (BZ) boundary by $\underline{K}$, the ratio of the mixing coefficients is $C_{\underline{K}}/C_{\underline{K},-\underline{G}} = \pm 1$. It is a well known result that the normal state energy has two roots in such a case.

We expect that some similar splitting may also be observed in the energy gap parameter $\Delta_{\underline{k}}$ when $\underline{k}$ lies on a BZ boundary. In a two plane wave region, the pairing interaction $V_{\underline{k},\underline{k}'}$ becomes

$$-V_{\underline{K},\underline{k}'} = \{\tfrac{1}{2}V(0) + V_{\underline{k}'}\} \pm V_{\underline{k}'}(\underline{G}) \tag{18}$$

where

$$V_{\underline{k}'}(\underline{G}) = \tfrac{1}{2}[V(\underline{G}) + V(-\underline{G}) + \sum_{\underline{g}}{}' U(\underline{g})\{V(\underline{g}+\underline{G}) + V(\underline{g}-\underline{G})\} \chi_{\underline{k}',\underline{g}}] \tag{19}$$

and $V_{\underline{k}'}$ is defined by Eq.(12). As we have previously discussed the main contribution to the sum over $\underline{k}'$ in Eq.(1) is from regions away from BZ boundaries. Thus, in calculating $V_{\underline{K},\underline{k}'}$, we must still use Eq.(18) for the wavefunction $\phi_{\underline{k}'}(\underline{r})$ in Eq.(7).

The gap equation now has a solution of the form

$$\Delta_{\underline{K}} = \Delta \pm \Delta_{\underline{G}} \tag{20}$$

where

$$\Delta = \sum_{\underline{k}'} \{\tfrac{1}{2}V(0) + V_{\underline{k}'}\} \frac{\Delta_{\underline{k}'}}{2E_{\underline{k}'}} \tanh(\tfrac{1}{2}\beta E_{\underline{k}'}), \quad \Delta_{\underline{G}} = \sum_{\underline{k}'} V_{\underline{k}'}(\underline{G}) \frac{\Delta_{\underline{k}'}}{2E_{\underline{k}'}} \tanh(\tfrac{1}{2}\beta E_{\underline{k}'}). \tag{21}$$

Thus, the energy gap parameter is split at the zone boundary as a result of the crystal structure.

It is now convenient to summarize the results for the gap $\Delta_{\underline{k}}$ using our modified Gor'kov potential and NFE Bloch functions:

$$\Delta_{\underline{k}} = <\Delta>(1 + a_{\underline{k}}) \tag{22}$$

for $\underline{k}$ away from Brillouin zone boundaries, and

$$\Delta_{\underline{K}} = \Delta \pm \Delta_{\underline{G}} \tag{23}$$

when $\underline{K}$ lies on a BZ boundary in a two plane wave region. In the following section we will be concerned only with Eq.(22).

3. Comparison With Gallium Tunneling Data

In order to calculate the anisotropy function $a_{\underline{k}}$, as given by Eq.(15), we need the three quantities $\chi_{\underline{k},\underline{g}}$, $U(\underline{g})$, and $V(\underline{g})/V(0)$. We will discuss each of these in turn.

The quantity $\chi_{\underline{k},\underline{g}} - <\chi_{\underline{g}}>$ in Eq.(15) contains the only $\underline{k}$ dependence in our expression for $a_{\underline{k}}$. Clearly, it is singular if $g = 2k_F$ or if $\varepsilon_{\underline{k}} = \varepsilon_{\underline{k}\pm\underline{g}}$. The approach of a singularity of the last type is an indication of where our expression for the anisotropy function is not valid. Singularities in $\chi_{\underline{k},\underline{g}}$ corresponding to some simple Fermi surface-Brillouin zone intersections are shown in Fig.1.

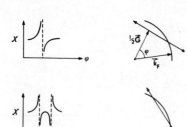

Fig.1:
Examples of singularities in the quantity $X_{\underline{k},\underline{g}}$ corresponding to some simple Fermi surface-Brillouin zone intersections.

The normal state pseudopotential is conventionally assumed to be a sum of spherical ionic terms

$$U(\underline{r}) = \sum_n u(\underline{r} - \underline{R}_n) \tag{24}$$

where n sums over all ions in the crystal. Then, the pseudopotential $U(\underline{g})$ becomes $S(\underline{g})u(\underline{g})$ where $S(\underline{g})$ is the structure factor,

$$S(\underline{g}) = (1/N) \sum_n e^{i\underline{g}\cdot\underline{R}_n}, \tag{25}$$

and $u(g)$ is the Abarenkov-Heine /9/ ionic pseudopotential form factor for scattering at the Fermi surface /10/. In the calculation of $a_{\underline{k}}$ to follow, we will use the analytical form of $u(g)$ due to Veljković and Slavić /11/.

The only unknown in the NFE formula for $a_{\underline{k}}$ is the average pairing interaction $V(\underline{g})$. This quantity appears in Eq.(15) for the anisotropy function only in the reduced form $V(\underline{g})/V(0)$. Until a microscopic theory of $V(\underline{r})$, and hence $V(\underline{g})$, is available we will have to content ourselves with a semi-empirical approach. In analogy to $U(\underline{r})$ we will represent the average pairing potential $V(\underline{r})$ by

$$V(\underline{r}) = \sum_n v(\underline{r} - \underline{R}_n) \tag{26}$$

where v is taken to be spherically symmetric about each ion position
Then, $V(\underline{g})$ also has the separable form $S(\underline{g})v(g)$.

Ideally, we would now like to treat $<\Delta>$ and each $v(g)/v(0)$ in the sum in Eq.(15) as an adjustable parameter in fitting to the gallium energy gap data /2/. There are, unfortunately, an unworkable number of terms contributing to the sum. Thus, we were forced to first define some simple model for $v(g)$. It was found that a simple rectangular cut-off model provided an excellent fit to the data. Fig.2 shows a comparison of the theory to the data for a $v(g)/v(0)$ which is small in the region $g \lesssim 4k_F$ and takes on a large negative value between $4k_F \lesssim g \lesssim 5k_F$.

The empirical $v(g)$ used implies a form for $v(r)$ which is shown in Fig.3. This form is indeed physically interesting in that it predicts a repulsive interaction ($v(r) < 0$) within the Ga^{3+} ion core radius, $r < r_c$, and an attractive interaction ($v(r) > 0$) just beyond r_c or in the immediate interstitial region.

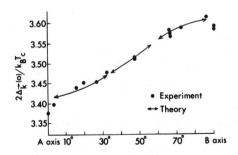

Fig.2: Comparison of the theory to experimental gap data in the gallium AB plane taken from Ref./2/. Breaks in the theory line indicate the regions of invalidity of the NFE perturbation model

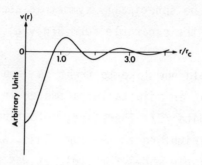

Fig.3: The form of the ionic term v(r) for gallium

The authors would like to thank W.D.Gregory and W.N.Mathews, Jr., for many helpful discussions. We are especially grateful to WDG for his help in the presentation of this paper in Vienna.

References

/1/ A.G.Shepelev, Soviet Phys.-Uspekhi 11, 690 (1969)
/2/ W.D.Gregory, A.J.Grekas, S.Horn, L.Morelli, paper C-14, this volume, p.265
/3/ J.Bardeen, L.N.Cooper, J.R.Schrieffer, Phys.Rev. 108, 1175 (1957)
/4/ D.Markowitz, L.Kadanoff, Phys.Rev. 131, 563 (1963)
/5/ J.R.Clem, Annals of Phys. 40, 268 (1966)
/6/ A.J.Bennett, Phys.Rev. 140, 1902 (1965)
/7/ C.R.Leavens, J.P.Carbotte, Annals of Phys. 70, 338 (1972)
/8/ L.P.Gor'kov, Soviet Phys.-JETP 7, 505 (1958)
/9/ I.V.Abarenkov, V.Heine, Phil.Mag. 12, 1249 (1965)
/10/ Strictly speaking, u(g) is a matrix element of a non-local operator u between the states $\underline{k}$ and $\underline{k}+\underline{g}$. It becomes a function of g only in the "on-Fermi-sphere" approximation. See Sections 2, 11, 12, 14 and 21 of the M.L.Cohen and V.Heine article in H.Ehrenreich, F.Seitz, and D.Turnbull, ed., Solid State Physics, Vol.24 (Academic Press, New York, 1970)
/11/ V.Veljkovič, I.Slavič, Phys.Rev.Lett. 29, 105 (1972).

C-16 SPECIFIC HEAT OF SUPERCONDUCTING ZINC*

R.W. Stark and S. Auluck

Department of Physics, University of Arizona

Tucson, Arizona 85721, U.S.A.

One of the experimental manifestations of superconductivity that has not yet received a quantitative understanding is the electronic specific heat C_{es} in most elemental superconductors. Many of these exhibit significant deviations from the BCS prediction usually yielding a larger value of C_{es} than that predicted for $T \ll T_c$. For $T_c/T \gtrsim 2$ the BCS model gives $C_{es}(T)/\gamma \cdot T_c = 8.5 \exp(-1.44\, T_c/T)$ for an isotropic ($T=0$) energy gap $\Delta_0 = 1.76\, k_B T_c$. The slope of $C_{es}(T)$ is 1.44 instead of 1.76 because of two significant contributions to the energy change with decreasing temperature: one of these is the condensation of further Cooper pairs into the ground state, while the other comes from the increase in the energy gap. It is traditional to interpret this type of deviation as arising from anisotropy in the energy gap Δ_k and estimates of this anisotropy, for simple metals, have been based on the C_{es} data. Although independent confirmation of this approach has been generally lacking, we will demonstrate in the following, that such confirmation exists for zinc, which shows strong deviations from the BCS behavior.

We have calculated the zinc band structure and Fermi surface

---

*Work supported by the National Science Foundation.

using a nonlocal pseudopotential model based on that given by Stark and Falicov /1/; this model agrees very well with all known data on the geometry of the Fermi surface /2/. In addition we have determined the many-body mass renormalization parameter λ_k (and in particular its anisotropy) by detailed comparison of calculated and measured cyclotron masses /3/. Though significant anisotropy exists in λ_k, it has a particularly simple form; λ_k is constant for each Fermi surface sheet but differs from sheet to sheet. The dominant electron-phonon part of λ_k resulting from our calculation for each sheet is: $\lambda^l = 0.47$ for the third band lens, $\lambda^m = 0.36$ for the second band monster, and $\lambda^c = 0.23$ for the first band cap. The corresponding fraction of the total renormalized density of states for each of these three sheets is: 0.25 for the lens, 0.70 for the monster and 0.05 for the cap.

Since the diagonal electron-phonon interactions, which determine λ_k, are essentially the same as the off-diagonal electron-phonon interactions determining Δ_k, we expect from these constant λ^i that, in the superconducting state, each sheet of the Fermi surface will be characterized by a different constant Δ^i at $T = 0$. Using a simple anisotropic generalization of the Eliashberg equations and our values for λ^i we predict the ratios

$$\Delta^l : \Delta^m : \Delta^c = 2.30 : 1.76 : 1.0. \tag{1}$$

Our ratio $\Delta^l/\Delta^m = 1.31$ agrees very well with the ratio $\Delta^l/\Delta^m = 1.29$ obtained experimentally from microwave absorption data /4/. Using the microwave absorption data to obtain absolute values we find $\Delta^l = 2.0 \, k_B T_c$, $\Delta^m = 1.52 \, k_B T_c$ and $\Delta^c = 0.87 \, k_B T_c$. Thus it is apparent that the one gap corresponding to only 5% of the renormalized density of states is much smaller than the other two; it is so small that few Cooper pairs will have condensed to the ground state for T near T_c. However, since the electron-phonon interactions in zinc are not extremely anisotropic and since 95% of the intermediate states

Specific Heat of Superconducting Zinc

are on regions of the Fermi surface with significantly larger energy gaps, we can anticipate that $\Delta^c(T)$ will closely follow $\Delta^l(T)$ and $\Delta^m(T)$ which in turn will be quite similar to the BCS variation. The mean of the large gaps

$$\frac{0.25\,\Delta^l + 0.70\,\Delta^m}{0.95} = 1.65 \tag{2}$$

is close to the BCS value of 1.76. Since only 5% of the intermediate states for these will be on the cap (for which a significantly large fraction of the states will be blocked by thermal excitation of normal state electrons) we expect their contribution to C_{es} to be 95% of the BCS contribution. For the cap we expect an exponential contribution which is given approximately by $\exp(-0.71\,T_c/T)$, i.e., equivalent to the BCS slope of 1.44 instead of 1.76. Thus we predict from the normal state data that the form of $C_{es}(T)/\gamma.T_c$ for zinc should be:

$$\frac{C_{es}(T)}{\gamma T_c} \sim 0.95 \times C_{es}^{BCS}(T) + 0.05\,\exp(-0.71\,\frac{T_c}{T}). \tag{3}$$

Figure 1 shows this curve (AGM) together with the experimental data obtained by Ducla-Soares and Cheeke for zinc plotted on a semilog

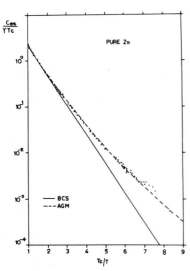

Fig.1:
The low temperature electronic specific heat $C_{es}(T)/\gamma T_c$ for superconducting zinc is plotted on a semilog scale vs T_c/T. The experimental data (points) are compared with the BCS theory and our anisotropic gap model discussed in the text.

scale versus T_c/T /5/. Also shown in the figure is $C_{es}^{BCS}(T)/\gamma T_c$, the BCS curve, which fits the data quite well near T_c. Note that the difference between the experimental data and $0.95 \times C_{es}^{BCS}(T)$ yields a straight line on the semilog plot with slope $-0.65\, T_c/T$. Also note that the intercept of this curve at $T = T_c$ is 0.035 of the extrapolated BCS curve of slope $-1.44\, T_c/T$. Thus the experimental data is given quite well by:

$$\frac{C_{es}^{exp}(T)}{\gamma T_c} \sim 0.95 \times C_{es}^{BCS}(T) + 0.035\, \exp(-0.65\, \frac{T_c}{T}) \qquad (4)$$

which is in remarkable agreement with the prediction of equation (3) based on normal state data. This agreement confirms the accepted belief, that the significant deviations of $C_{es}^{exp}(T)$ from $C_{es}^{BCS}(T)$ results from anisotropy in Δ_k. The most remarkable feature in this particular case is that the deviation results from only 5% of the total renormalized density of states for which Δ_k is significantly smaller than the mean $\overline{\Delta}$.

Thus we see that a detailed knowledge of the electronic properties in the normal state is sufficient to predict correctly the electronic properties in the superconducting state. The analysis in this paper supercedes the analysis of the specific heat data given earlier /3/.

Present address of S.Auluck: Department of Physics, Punjab Agricultural University, Ludhiana-141004, India.

References

/1/ R.W.Stark, L.M.Falicov, Phys.Rev.Lett. $\underline{19}$, 795 (1967)
/2/ R.W.Stark, S.Auluck, to be published
/3/ S.Auluck, J.Low Temp.Phys. $\underline{12}$, 601 (1973)
/4/ J.B.Evans, M.P.Garfunkel, D.A.Hays, Phys.Rev. $\underline{B1}$, 3629 (1970)
/5/ E.Ducla-Soares, J.D.N.Cheeke, in Proc. 12th Conf. on Low
     Temp.Phys., Kyoto (Japan), Academic Press of Japan
     (E.Kanda, Editor), p. 305, 1971

C-17 UNUSUAL RESISTANCE EFFECT SHOWN IN A PERIODIC S-N-S

SYSTEM (Pb-Sn LAMELLAR EUTECTIC)

J.M.Dupart, J.Rosenblatt [+], J.Baixeras

Laboratoire /1/ de Génie Electrique des Universités de

Paris VI et Paris XI, F-92260 Fontenay-aux-Roses

The two component Pb-Sn lamellar eutectic systems obtained by directional solidification in the form of 20 μm thick ribbons, if sufficiently regular, are suitable for studying S-N-S junctions in series /2/.

For a current perpendicular to the lamellae we have measured the resistivity $\rho_\perp(T)$ of our samples. $\rho_\perp(300\ K)$ has about the same value for all samples ($\simeq 5 \times 10^{-7}\ \Omega m$). The ratio $\rho_\perp(300\ K)/\rho_\perp(8\ K)$ can reach the value 10^5 (Table 1); the values of the mean free path then calculated /3/ are quite large (Table 1), showing the good crystallographic quality of the lamellae. In perfect lamellar areas, Pb and Sn lamellae are indeed monocrystals /4/; the Sn crystal is polygonized whereas the Pb crystal is bent. Each Sn subgrain, in which the lamellar crystals are parallel within 5×10^{-4} rad, is 10 to 20 lamellae wide and 250 μm long. The relationships between the Sn and Pb phases have been determined /4/.

On these ribbons (about 100 junctions in series) with lamellar spacing q_E ranging from 3.6 μm to 8 μm, a non zero critical current

---

[+] and Institut National de Sciences Appliquées, F-35031 Rennes

I_c was found up to a temperature T*, $T_c(Sn) = 3.7$ K < T*, T* < $T_c(Pb)$ = 7.4 K (Table 1). The temperature and normal width ($a_N = 2/3\ a_E$) dependence of the critical current can be explained within the framework of Bardeen and Johnson calculations /5/.

Table 1

Sample No.	1	2	3	4
a_E (μm)	3.6	4.9	6.7	8.1
T*(K)	7.2	6.7	5.7	4.7
$\rho_\perp$ (Ωm) (T = 8 K)	6.9×10^{-9}	1.7×10^{-12}	2×10^{-11}	5.5×10^{-9}
ℓ (μm) (T = 8 K)	0.2	600	50	0.2
$\rho_\perp$ (300 K)/$\rho_\perp$(8 K)	70	3×10^5	2.5×10^4	90

The zero field V(I) characteristics of all samples show a linear region, where $V \simeq R_D(I - I_c)$. In the temperature region, where experimentally $I_c \neq 0$, this always extends over a current interval such that $I - I_c \geq I_c$. The resulting differential resistivities $\rho_\perp$ are plotted as a function of the reduced temperature $t = T/T_c(Pb)$ for four samples in Fig.1. Samples 1 and 4 having similar (high) normal resistivity show a decrease of $\rho_\perp$ with temperature, while those (samples 2 and 3) having very low normal resistivity present a significant increase of $\rho_\perp$ just below T* (three orders of magnitude in one case).

Forward moving electrons in samples 1 and 4 suffer roughly 10 collisions per period at T = 8 K. This quantity decreases to about one collisions per 120 periods in the sample of lowest resistivity ($\rho_{\perp N} = 1.7 \times 10^{-12}$ Ωm). It seems natural to assume then, that the high resistivity value is mainly due to elastic scattering by defects or impurities, but that electron-phonon (inelastic) collisions are pre-

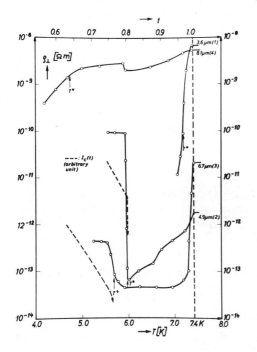

Fig.1: Differential resistivities $\rho_\perp$ and critical currents I_c as function of reduced temperature

dominant when $\rho_{\perp N} \simeq 10^{-12}$ Ωm. The behavior of single S-N-S contacts under conditions of rather strong interface scattering was studied by Pippard et al. /6/ and Harding et al. /7/. Their junctions show a steady decrease of resistance with temperature as our samples 1 and 4, which is explained as being due to electron hole reflection at the N-S boundaries /6/. We conclude, that a high resistivity alternating structure simply reproduces single contact behavior. In the case of low resistivity samples, the fact, that the dynamical resistance is a constant over a current interval of a few times the critical current itself, precludes an identification of the marked increase of $\rho_\perp$ at T* with the singularity of dV/dI at I $\gtrsim$ I_c in ideal Josephson junctions /8/.

Van Gelder /9/ has shown that a periodic system with $\Delta = 0$ and Δ_o in alternating regions of length a_N and a_S, respectively, will present a band structure in the elementary excitation spectrum. This result is obtained from a solution of Bogoliubov's equations employing periodic boundary conditions and assuming perfect matching of the wave functions at the S-N boundaries. The bands are defined by the following equations:

$$\cos\{(k_z - k_{zF})d\} = F(\varepsilon) \qquad (1)$$

$$F(\varepsilon) = \cos(A_N\varepsilon)\cos(A_S(\varepsilon^2-1)^{\frac{1}{2}}) - (\varepsilon/(\varepsilon^2-1)^{\frac{1}{2}})\sin(A_N\varepsilon)\sin(A_S(\varepsilon^2-1)^{\frac{1}{2}}) \qquad (2)$$

where k_z is the Bloch momentum in z-direction,

$$k_{zF}^2 = k_F^2 - k_x^2 - k_y^2; \quad A_{N,S} = a_{N,S}\, m\Delta_o/\hbar^2 k_{zF}$$

and $\varepsilon = E/\Delta_o$, E being the quasiparticle energy measured from the Fermi surface. Equation (2) holds for $\varepsilon > 1$. Imaginary arguments for $\varepsilon < 1$ result in a suitable replacement of the trigonometric by hyperbolic functions. For given values of A_N and A_S, which depend on temperature through $\Delta_o(T)$, the range $-1 < F(\varepsilon) < 1$ corresponds to allowed bands with $-\pi/d \leq q \leq \pi/d$, $q = k_z - k_{zF}$. Examples of such a band structure are given in Fig.2. We assume, that an excitation spectrum of this kind is indeed present in our very low resistivity samples. On the contrary, strong scattering N-S boundaries may well prevent

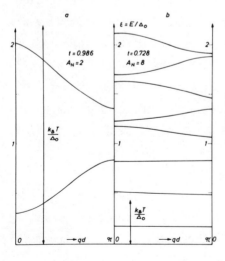

Fig.2:
Dispersion laws predicted by Eq.(2) for $k_{zF} = k_F$

matching of the wave functions resulting in the single junction behavior exhibited by the high resistivity samples.

The theory of supercurrent flow in pure S-N-S contacts has been discussed by Bardeen and Johnson /5/. Quasiparticles occupy discrete energy levels E_n in the pair potential well (N region) of depth Δ_0. An externally applied flow produces a displacement of the Fermi sea in k-space until it touches the first energy level. This determines the condition for maximum supercurrent at T = 0 K, with a phase difference of π. At $T \neq 0$ K the probability of occupation of different levels has to be taken into account giving a current density:

$$J = n_e e v_s + \frac{e}{m} \sum_{k_{zF} > 0} \frac{\hbar k_{zF}}{2d^*} \{f(E_n + \hbar v_s k_{zF}) - f(E_n - \hbar v_s k_{zF})\} \tag{3}$$

where v_s is the flow velocity, d^* the effective thickness of the N-region, n_e the electron density and f the Fermi-Dirac function. A similar expression should apply to our band system, provided that the sum is restricted to levels in allowed bands and d^* is replaced by $d = a_E$ to achieve a proper normalization of the wave functions.

Consider now relaxation of the electron distribution, when $I \gtrsim I_c$. At T close to T_c(Pb) the thermal smearing spans the whole band system (Fig.2a) and most of the electron phonon processes are allowed. Furthermore, I_c is practically zero and the values of f in (3) are close to the equilibrium values. The usual theory /10/ of the "ideal" resistivity of metals is applicable predicting a slow decrease of resistance with temperature, as observed with sample 2, masked perhaps in sample 3 by some elastic scattering. On the other hand, at temperatures where the bands are narrow enough to be considered as degenerate single levels with $E_n - E_{n-1} \gtrsim k_B T$ (Fig.2b) most electron transitions will be interband transitions. The collision term in the Boltzmann equation and, therefore, the measured resistance will be proportional to the difference $f_k - f_{k'}$ between the

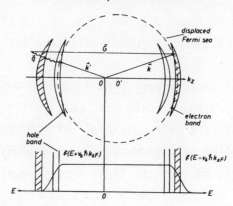

Fig.3: Umklapp phonon emission; $\vec{G}$ is a reciprocal lattice vector

initial and the final electron distributions /10/. It is important to notice, that f_k and $f_{k'}$ are <u>not</u> close to equilibrium at $I \gtrsim I_c$. Indeed, considering the umklapp process illustrated in Fig.3, one sees that a reasonable approximation to f_k is a displaced Fermi distribution, as in Eq.(3). Consequently $\rho_\perp$ and I_c should be strongly correlated, as is effectively observed in our case. Assuming equally spaced levels (corresponding approximately to our bands near T*) Bardeen and Johnson /4/ have calculated the sum

$$\sum_{k_{zF} > 0} \{f(E_n + \hbar v_s k_{zF}) - f(E_n - \hbar v_s k_{zF})\}$$

for single S-N-S junctions, which in our case should be proportional to $\rho_\perp$. One obtains

$$\rho_\perp \propto I_c \propto e^{-2\pi^2/x_F} \qquad (4)$$

with $x_F = 2 E_o(T)/k_B T$. The critical currents $I_c(t)$ of our low resistivity multijunction samples are plotted together with $\rho_\perp(t)$ in Fig.1. We can see that the increase of $\rho_\perp$ with decreasing T and the appearance of I_c are indeed coincident.

We are indebted to J.J.Favier and Dr.R.Racek (Ecole des Mines, Paris) for the sample preparation. We wish to acknowledge the invaluable cooperation of C.Bodin.

References

/1/ Laboratoire associé au CNRS No. 127
/2/ J.M.Dupart, J.Baixeras, Proc. LT 14 (Eds. M.Krusius and M.Vuorio), North-Holland, 1975, vol.4, p. 132
/3/ R.G.Chambers, Proc.Roy.Soc. $\underline{A215}$, 481 (1952)
/4/ B.Labulle, C.Petipas, J.Crystal Growth $\underline{28}$, 279 (1975)
/5/ J.Bardeen, J.L.Johnson, Phys.Rev. $\underline{B5}$, 72 (1972)
/6/ A.B.Pippard, J.G.Shepherd, D.A.Tindall, Proc.Roy.Soc.Lond. $\underline{A324}$, 17 (1971)
/7/ G.L.Harding, A.B.Pippard, J.R.Tomlinson, Proc.Roy.Soc.Lond. $\underline{A340}$, 1 (1974)
/8/ D.E.McCumber, J.Appl.Phys. $\underline{39}$, 3113 (1968)
/9/ A.P. Van Gelder, Phys.Rev. $\underline{181}$, 787 (1969)
/10/ J.M.Ziman, The Physics of Metals $\underline{1}$ (Cambridge University Press) 1969.

C-18   THE UPPER CRITICAL FIELD OF SUPERCONDUCTING

POLYSYLFUR NITRIDE, $(SN)_x$ (Abstract Only)

L.J.Azevedo, W.G.Clark, G.Deutscher

Department of Physics, University of California

Los Angeles, California 90024, U.S.A.

R.L.Greene, G.B.Street

IBM Research Laboratory, San Jose, California 95193, U.S.A.

L.J.Suter

Department of Physics, Stanford University

Stanford, California 94305, U.S.A.

A measurement of the temperature and angular dependence of the upper critical magnetic field H_{c2} in crystals of polymeric $(SN)_x$ is reported. A large anisotropy is observed at all temperatures with $H_{c2} = 8.1 \pm 0.4$ kOe parallel to, and $H_{c2} = 870 \pm 80$ Oe perpendicular to, the polymer axis at 0 K. The results are explained in terms of the polymeric crystal structure and fibrous morphology of $(SN)_x$.

LIST OF PARTICIPANTS

J.L.Bostock - Room 6-204, Massachusetts Institute of Technology, Cambridge, Mass. 02139, U.S.A.

J.P.Carbotte - Department of Physics, McMaster University, Hamilton, Ontario L8S 4M1, Canada

W.G.Clark - Physics Department, University of California, 405 Hilgard Ave, Los Angeles, Ca. 90024, U.S.A.

J.M.Dupart - Laboratoire de Génie Électrique des Universités Paris VI et XI, 33, Av. du Géneral Leclerc, F-92260 Fontenay-aux-Roses, France

P.Entel - Institut für Theoretische Physik, Universität Köln, D-5 Köln 41, Zülpicherstrasse 77, W-Germany

R.W.Genberg - Department of Physics, Adelphi University, Garden City, Long Island, N.Y. 11530, U.S.A.

C.E.Gough - Department of Physics, University of Birmingham, Birmingham B15 2TT, U.K.

W.D.Gregory - Department of Physics, Georgetown University, Washington, D.C. 20057, U.S.A.

H.Haslacher - Elin-Union, Penzingerstrasse 76, A-1141 Wien, Austria

O.Hittmair - Institut für Theoretische Physik I, Technische Universität Wien, Karlsplatz 13, A-1040 Wien, Austria

R.P.Huebener - Lehrstuhl für Experimentalphysik II, Universität Tübingen, D-74 Tübingen 1, Morgenstelle, W-Germany

H.Kiessig - Institut für Physik am Max-Planck-Institut für Metallforschung, D-7 Stuttgart 80, Büsnauerstrasse 171, W-Germany

Participants

H. Kirchmayr - Institut für Experimentalphysik, Technische Universität Wien, Karlsplatz 13, A-1040 Wien, Austria

R. Leitner - Institut für Experimentalphysik, Technische Universität Graz, Rechbauerstrasse 12, A-8010 Graz, Austria

M.L.A. MacVicar - Room 8-201, Massachusetts Institute of Technology, Cambridge, Mass. 02139, U.S.A.

F.K. Mullen - Department of Physics, Adelphi University, Garden City, Long Island, N.Y. 11530, U.S.A.

H. Novotny - Institut für Theoretische Physik I, Technische Universität Wien, Karlsplatz 13, A-1040 Wien, Austria

B. Obst - Institut für Experimentelle Kernphysik, Kernforschungszentrum Karlsruhe, D-75 Karlsruhe, Postfach 3640, W-Germany

T. Ohtsuka - Department of Physics, Tohoku University, Sendai 980, Japan

H.R. Ott - Laboratorium für Festkörperphysik, CH-8049 Zürich, Hönggerberg, Switzerland

E. Paumier - Laboratoire de Physique du Solide, Université de Caen, F-14032 Caen Cedex, France

W. Rodewald - Optisches Institut, Technische Universität Berlin, D-1 Berlin 12, Strasse des 17.Juni 135, W-Germany

E. Schachinger - Institut für Theoretische Physik und Reaktorphysik, Technische Universität Graz, Steyrergasse 17, A-8010 Graz, Austria

J. Schelten - Institut für Festkörperforschung, Kernforschungsanlage Jülich, D-517 Jülich, Postfach 365, W-Germany

R. Schneider - Institut für Festkörperforschung, Kernforschungsanlage Jülich, D-517 Jülich, Postfach 365, W-Germany

Participants

E.Seidl - Atominstitut der Österreichischen Universitäten,
Schüttelstrasse 115, A-1020 Wien, Austria

J.Sporna - Atominstitut der Österreichischen Universitäten,
Schüttelstrasse 115, A-1020 Wien, Austria

K.Takanaka - Department of Engineering Science, Tohoku University,
Sendai 980, Japan

H.Teichler - Institut für Physik am Max-Planck-Institut für Metallforschung, D-7 Stuttgart 80, Büsnauerstrasse 171, W-Germany

H.Ullmaier - Institut für Festkörperforschung, Kernforschungsanlage Jülich, D-517 Jülich, Postfach 365, W-Germany

H.W.Weber - Atominstitut der Österreichischen Universitäten,
Schüttelstrasse 115, A-1020 Wien, Austria.

AUTHOR INDEX

Auluck, S.	293
Azevedo, L.J.	307
Baixeras, J.	299
Bostock, J.L.	213, 257
Carbotte, J.P.	183
Cirillo, N.C. jr.	283
Clark, W.G.	307
Clinton, W.L.	283
Deutscher, G.	307
Dupart, J.M.	299
Entel, P.	47
Essmann, U.	69
Farrell, D.E.	165
Genberg, R.W.	171
Gough, C.E.	79
Greene, R.L.	307
Gregory, W.D.	265
Grekas, A.J.	265
Heiden, C.	177
Hembach, R.J.	171
Horn, S.	265
Huebener, R.P.	165
Kampwirth, R.T.	165
Kiessig, H.	69
Mac Vicar, M.L.A.	213, 257
Milkove, K.R.	257
Morelli, L.	265
Mullen, F.K.	171
Obst, B.	139
Ohtsuka, T.	27
Ott, H.R.	87
Peter, H.	47
Rodewald, W.	159
Rosenblatt, J.	299
Schelten, J.	113, 177
Schneider, R.	177
Seidl, E.	57
Stark, R.W.	293
Street, G.B.	307
Suter, L.J.	307
Takanaka, K.	75, 93
Teichler, H.	7, 65, 69
Weber, H.W.	57
Wiethaup, W.	69

SUBJECT INDEX

(Page numbers refer to beginning of relevant paper)

Average upper critical field 27, 47, 75, 93

B_0 7, 69, 79, 139
Boundary scattering experiments 213, 265

Correlations between crystal lattice
- and domain structure 139, 159, 165
- and flux line lattice 7, 65, 93, 113, 139, 159
- and flux spots 139, 159, 165
- and superlattice 139

Cubic materials 7, 27, 57, 65, 69, 93, 159

Decoration technique 79, 113, 139, 159
Domain structure 139, 159, 165

Electromagnetic radiation 213
Electron band structure 7, 47, 65, 87, 183, 213, 293
Electron-phonon interaction 7, 27, 47, 57, 65, 87, 183, 213, 265, 283
Eliashberg equations 183, 283, 293
Energy gap 7, 57, 65, 93, 113, 183, 213, 257, 265, 283, 293

Flux line lattice 7, 93, 113, 139, 159

Flux spot 139, 159, 165

H_c 69, 213
H_{c1} 7, 27, 69, 93
H_{c2} 7, 27, 47, 57, 65, 69, 75, 79, 307
H_{c2} at very low temperatures 27, 75

Impurity dependence
- of energy gap 213
- of flux line lattice morphology 113
- of H_{c2} anisotropy 7, 27, 57, 65

Impurity scattering 213, 265
Intermediate mixed state 27, 79, 139
Intermediate state 159, 165

Josephson tunneling 299

κ_2 7, 69, 79

Lamellar superconductors 299
Layered materials 7, 93
Low dimensional superconductors 307

Magnetization 79, 93, 171, 177
Magnetoelastic interaction 93, 113, 139
Markowitz-Kadanoff theory 7, 213, 267, 283

Microscopic anisotropy parameters 7, 57, 65
Microscopic gap calculations 183, 283
Multiple energy gap 213

Nearly free electron model 265, 283
Neutron diffraction 57, 79, 113
Nonlocal theory 7, 27, 57, 65, 93, 139

Phonon spectrum 7, 87, 139, 183, 213
Pinning effects
 - bulk pinning 57, 79, 113, 139, 159
 - surface pinning 79
Polysulfur nitride 307

Quasilocal theory 7, 27, 75, 93, 113, 139, 171, 177

S-N-S junctions 299
Specific heat 79, 213, 293
Stress dependence of H_c, T_c 87
Superlattice 139
Surface energy 27, 139

Temperature dependence of H_{c2} anisotropy 7, 27, 57, 65, 69, 75, 93
Thermal conductivity 79, 139, 213
Torque measurements 93, 171, 177
Tunneling experiments 183, 213, 257, 265, 283
Two band model 47
Type-I - type-II transition 27, 69

Ultrasonic attenuation 79, 213
Uniaxial superconductors 7, 87, 93